Business Computer Systems Design

Business Computer Systems Design

Kathleen A. Dolan

MITCHELL PUBLISHING, INC.
Santa Cruz, California

To Carla

Production: Hal Lockwood, Bookman Productions
Production coordination: Susan Defosset
Cover design: Steve Renick
Text design: Jeanne Tennant
Manuscript editor: Larry McCombs
Illustrations: Carl Brown
Composition: Graphic Typesetting Service

Cover photos clockwise from the
top courtesy of American Airlines,
Anacomp, Inc., AT&T, Anacomp, Inc.,
and General Instruments Corp.

Library of Congress Cataloging in Publication Data

Dolan, Kathleen, 1950–
   Business computer systems design.

   Bibliography: p.
   Includes index.
   1. System design.   2. System analysis.   3. Business—
Data processing.   I. Title.
QA76.9..S88D65   1984   001.64   83–25156
ISBN 0-938188-20-8 (pbk.)

Printed in the United States of America

10   9   8   7   6   5   4   3   2   1

Contents

PART 1

Designing Programs

PART 2

Other System Components

PART 3

After Design

Preface

Business Computer Systems Design was written to fill the need for a college-level textbook that not only discusses the theories of program and system design but also presents design techniques that are currently used in the information systems industry. Whereas many college textbooks remain at the theoretical level, and others, geared primarily for the "computer systems professional," make large assumptions about the reader's knowledge and experience, *Business Computer Systems Design* recognizes the college student's need for theory backed by numerous realistic examples and illustrations.

Perhaps the most striking feature of *Business Computer Systems Design* is the extensive use of illustrations. There are over 175 of them in the text and appendix. Figures and diagrams are an important aspect of program and system design, so care has been taken to present a sample large enough to provide the student a model upon which to base his or her own design documents.

The system analysis and program design approach adopted in this book is based on the "design by dataflow" techniques developed by such pioneers as Larry Constantine, Tom DeMarco, Meilir Page-Jones, Wayne Stevens, and Edward Yourdon, to name a few. Those individuals and others have written extensively on the topic, and I heartily recommend their work to anyone planning to pursue system and program design as a career.

Many design textbooks present only program design, to the exclusion of other very important aspects of system design. More and more, the information systems industry is coming to the realization that well-designed programs do not a good computer system make. Attention must also be paid to the careful design of procedures, documentation, screen displays, error messages, printed reports, data storage, and employee training programs. Program design is certainly a major part of system design, but it is by no means the only part. In response to the real need for students now to be able to do more after graduation than code programs, *Business Computer Systems Design* also addresses the activities of designing data files and databases, capturing and presenting data, establishing and documenting procedures, and preparing training programs. The text is organized as follows.

Chapter 1 contains background information about computer system development and system specification documents, and it introduces a case study that will be used throughout the text for examples and exercises.

Part 1 Designing Programs

Chapters 2–6 present program design. These chapters explain the qualities of a good program, how to go about the program design process, and how to check your work in order to catch errors before continuing.

Part 2 Other System Components

Chapters 7 and 8 address the data component of a system—how to store data in files and databases, how to capture input data online and in batches, and how to present data on CRTs and in printed reports.

Chapter 9 examines procedures—how to establish them and how to document them in manuals and with screen displays.

Chapter 10 presents topics that enable the system developer to design training for the individual(s) who need to know how to use the system.

Part 3 After Design

Chapter 11 is an overview of system implementation. Students need to know what happens after design so they'll understand how crucial good design is to the successful development of a computer system.

Each chapter begins with a list of objectives and concludes with a chapter summary, a list of key words and definitions introduced in the chapter, and a series of exercises. Answers to questions and solutions to exercises can be found in the instructor's guide that accompanies *Business Computer Systems Design*.

Business Computer Systems Design is intended to be the primary text in a course on program design and in the system design part of a system analysis and design course sequence, such as the CIS-5 course described in the DPMA's model Computer Information Systems curriculum. *Business Computer Systems Design* is also a valuable supplementary text in any advanced high-level language course, such as COBOL or Pascal.

The design techniques described in this text are language-independent, of course, and as such are presented without regard to any particular language implementation. In a few instances, where it seemed appropriate to present a short program or subprogram in order to illustrate the implementation of some design technique, COBOL was used as the programming language. The programs are merely examples, and no assumptions were made about the reader's experience with COBOL—in fact, COBOL was selected because of the relative ease with which it can be read and understood, even by a non-COBOL-programmer. Every COBOL example can easily be changed to BASIC, PL/I, Pascal, or FORTRAN without losing the illustrative benefits. Space constraints, however, do not permit us to do so, and it would serve no purpose other than to compare programming languages—but this is a book on design.

In the course of preparing the manuscript for *Business Computer Systems Design,* versions of it passed under the critical eyes and red pencils of a number of people to whom I am indebted. Without their mostly gentle and always honest criticisms and suggestions, this text would not be in your hands today. Therefore I wish to thank the many college teachers who have taken the system development course during the National Computer Educators' Institute, both at James Madison University (Virginia) and at Central State University (Oklahoma). *Business Computer Systems Design* is a direct by-product of that course. I extend my humble appreciation to Marilyn Bohl of IBM Corporation for her detailed criticisms and suggestions; to Vivie Babb of the State University of New York for her careful editing and reviews; and to Dave Kroenke, consultant and author, for the puffins when I needed them most. Finally, I owe a deep thanks to Carla Wirzburger and to my family for their boundless patience and support.

CHAPTER 1

Computer System Development: An Introduction

When you finish this chapter, you will be able to—

- name the five components of a computer system
- describe the four steps in the development of a computer system
- identify the three documents in the system specification
- describe the relationship between the dataflow diagram and process specifications
- describe the relationship between the data dictionary and dataflow diagrams
- interpret data dictionary entries
- interpret a leveled set of dataflow diagrams
- identify functional primitives on a dataflow diagram

INTRODUCTION

This chapter is divided into three parts. The first part describes the five components of a business computer system and the four-step system-development process. The second part introduces a fictitious case study used to illustrate and develop ideas and design techniques as they are presented. The case study is set in a public agency that serves the needs of blind citizens. The third part of this chapter describes the documents we use to specify the requirements of a system: dataflow diagrams, the data dictionary, and process specifications.

THE FIVE-COMPONENT MODEL

A business computer system is made up of five interfacing components: hardware, programs, data, procedures, and people. *Hardware* is the computer equipment itself: processors, printers, disk and tape drives, modems, video display terminals, and so forth. *Programs* are instructions for the computer hardware, written by programmers in computer languages such as Pascal, FORTRAN, COBOL, and BASIC. A program is like a recipe. *Data* are to programs what ingredients are to recipes: they are the raw facts stored in the computer and processed according to the instructions in the programs. Many *people* are involved in computer systems: some people run and service the hardware, some people write programs, and other people collect and prepare the data that go in and distribute the information that comes out to the people who ultimately use it. All of these people must know how to play their parts in the computer system. They follow *procedures,* or instructions that direct their interactions with the computer.

Any computer system, regardless of its size, has the five components. Think about a small system used by a plumbing-supplies distributor for bookkeeping and inventory applications. It might consist of an inexpensive microcomputer (hardware) running purchased applications packages (programs) to maintain the company's customer and inventory records (data), and it might be operated by the company's bookkeeper (people), who follows the directions in the reference manuals that came with the computer and software, as well as other company policies (procedures).

Now imagine a computer system used by a large international financial organization. Millions of dollars are invested in elaborate configuratons of computer hardware, from multiple processors to laser printers to satellite communications equipment to rooms full of disks upon which are stored billions of pieces of data. Huge sums of money are spent each year on data-processing specialists who write programs, run and service equipment, prepare data for input, and train company employees in computer-related areas.

Despite vast differences between the sizes and costs of the systems that provide services to the small plumbing-equipment distributor and to the international bank, each system is composed of the five components: hardware, programs, data, procedures and people.

THE SYSTEM-DEVELOPMENT PROCESS

Computer systems are developed in much the same way as products of other engineering fields. First, someone establishes the need for a new product to be built. This usually happens because someone perceives a problem: a difference between the way things are now and the way that they ought to be. This individual defines what the new product must do—that is, the requirements for the new product.

Next, specialists propose various alternative solutions to the problem. The alternatives are studied, modified, and verified. Finally, one alternative is chosen to be further developed.

The third step is called design. Models of the new product are built and subjected to tests that simulate as closely as possible the conditions under which the real product will have to function. The model, once refined, is used to guide implementation—the process of building and testing the real product. Finally, the product is put on the market for sale, or it is placed into operation.

This procedure works equally well for developing ballpoint pens, office buildings, pleasure boats, jet engines, running shoes, and computer systems. Define the problem and the ultimate requirements of the final product; develop alternative solutions for evaluation and selection; design a model of the system; then build it, test it, and implement it. Let us take a closer look at the development of computer systems.

SPECIFY SYSTEM REQUIREMENTS

We refer to the individual or individuals who need the computer system as the *user* (we use the singular, although there may be many individuals involved). The user, simply stated, has a problem and needs a solution. A data-processing specialist called a *system analyst,* or simply an *analyst,* works with the user to define precisely what the user's requirements are: what the system must do for the user, and under what constraints it must operate. One typical goal is that the computer will automatically perform some or all of the user's business activities; therefore, the analyst and user carefully document the user's business activities, as well as results produced by one activity that are either used in another activity or trigger execution of another activity. For example, a single sale to a customer can cause a statement to be sent to that customer at the end of the month, can cause a salesperson's commission to be increased, can decrease the company's inventory, and can cause an order to be placed to replenish depleted stock, resulting in a bill that must be paid to a vendor. All of these business activities affect one another, and all must be included in the system specification.

Having built a model of the enterprise's business activities, the user and the analyst can decide which business activities, if any, will be automated. Of course, the new system may also include features and functions that do not exist in the user's current system: these, too, are included in the system specification. The system specification becomes the target for the remaining steps in the development of the computer system.

PROPOSE ALTERNATIVES, EVALUATE THEM, SELECT ONE

Once the analyst knows exactly what the system must do for the user, he can begin to rough out some ideas for solving the problem. He develops several proposals, involving different pieces of computer hardware, different data-access methods, different applications programs, and requiring different numbers of people who follow different procedures. Each of the proposals will result in some measurable benefits to the user, and each will cost some amount of time and money to produce. The user weighs the potential benefits and likely costs of each of the proposals and selects the one that best suits the user's needs and resources.

DESIGN THE SYSTEM

Construction workers do not erect buildings based on the architect's watercolor rendition of the outside shell. Rather, they follow blueprints—excruciatingly detailed pictures of all the interfacing pieces of the structure. Drafting blueprints, although it requires an expenditure of both time and money, is a very necessary step. Few building purchasers would want an architect to skip this step and begin pouring foundations without detailed building plans.

Similarly, the system specification, like a watercolor painting of a building, defines what the user wants the system to look like. It is not intended to be the document that guides the process of building a system. Instead, it is used as the basis from which system blueprints will be developed. The detailed blueprints, or system-design documents, are needed by the people who will write programs, enter data, install and operate equipment, create files, and so on.

Each component of the new computer system must be carefully planned and designed. Programming specifications are written; databases and files and reports are designed; user and operator reference manuals are outlined; procedures are planned; training objectives are established and courses are outlined and even scheduled. All this is done before the actual product is built.

IMPLEMENT THE SYSTEM

Implementation of a computer system involves the coordinated efforts of many people: hardware specialists install equipment, programmers write and test programs, database experts build files and databases, technical writers produce reference manuals, and professional trainers teach users how to use the new system. Often these activities occur in parallel. Thus, at some point, all the pieces of the system must be brought together to run as a unit for the first time.

It is at this point that the quality of the design effort becomes evident. With careful design, implementation can be a smooth step; however, if the design step was sloppy, implementation will be a source of headaches and aggravation for implementors and users alike. There cannot be too much emphasis on design, in spite of the amount of time it requires. Often implementors want to "get on with it" and start building; users clamor for tangible results; and everybody seems to want to see code being written. However, the meaning of the old adage lingers: If you don't have time to do it right, you certainly don't have time to do it over.

THE FOUR STEPS OF SYSTEM DEVELOPMENT

Let us summarize. In the first step (specify the system requirements), we define *what* has to be done. In the second step (evaluate alternatives), we *select* the solution that best fits the user's budget and schedule. In the third step (design), we decide *how* the system will be built and tested. And in the fourth and final step (implementation), we *build and test* the system. All five components of a computer system are addressed at each step in the development process.

This text addresses the third step in the development of a computer system: design. Our goal is to produce detailed plans for the implementors, so they will build a system that satisfies the requirements agreed upon by the user and analyst.

We begin by examining an enterprise that perceived a problem and decided that computerizing part of its operations would help solve the problem. The Agency for the Blind is described in terms of its overall functions and problems. A system specification is presented because the techniques presented in subsequent chapters use some or all of these documents as their input.

THE STATE AGENCY FOR THE BLIND

The State Agency for the Blind serves the special needs of visually handicapped residents of the state. Its three divisions address different aspects of service: the Division of Children's Services provides for the special needs of children; the Division of Vocational Rehabilitation addresses postsecondary education and vocational training for its clients; and the Division of Adult Services helps individuals acquire independent living skills.

FINANCIAL OPERATIONS

Each division administers its own budget, and there is little direct involvement of one division with another, even though a client might be receiving services from more than one division at one time.

There is a fourth autonomous entity within the agency called the Business Office. The Business Office maintains financial records for the entire agency, and it communicates with the State Comptroller's Office in all matters of expenditures. No checks are ever issued directly by the agency: all expenditures are paid by the State Comptroller, miles away from the agency.

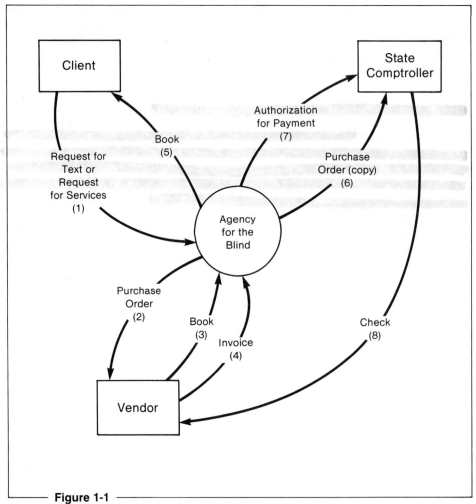

Figure 1-1

Flow of financial data in Agency for the Blind.

An example will help you understand how this works (see Figure 1-1). Let us presume that a client in high school (1) requests a particular textbook in large print. The agency issues (2) a Purchase Order and sends it to the vendor. The vendor fills the order, sending (3) the large-print Book and (4) an Invoice to the agency. The agency issues (5) the Book to the client and sends (6) a copy of the original Purchase Order and (7) an Authorization for Payment to the State Comptroller. The comptroller prepares and mails (8) a Check for the authorized amount to the vendor. The client gets his book from the agency, and the vendor gets his money from the comptroller. But where did the comptroller get the money?

Once each year the federal and state governments allocate funds for use on behalf of visually handicapped people. The agency, which administers these funds, never sees the actual money. Instead, the State Comptroller establishes accounts from which the funds are drawn as they are needed. The Business Office in the agency knows the beginning balances of all the accounts, and each Division Director knows how much money is available at the beginning of the year in each of his division's accounts.

CLIENTS

In order to receive services from the agency, one must be a visually handicapped resident of the state. Visually handicapped people are either legally blind or visually impaired.

A legally blind person either has no better than 20/200 vision in his better eye with correction, or has a field of vision no greater than 15°.

A visually impaired person has no better than 20/70 vision in his better eye with correction.

A client can be either a child or an adult. An adult is anyone 18 years old or over.

The term *schoolage* refers to a client who is under 21 years of age and has not yet graduated from high school.

The term *multihandicapped* refers to a client who, in addition to visual loss, also has a hearing loss, a neurological problem, or is otherwise physically or mentally challenged.

Combinations of age, educational level, extent of visual handicaps, and physical and emotional problems determine what funds, if any, can be spent on behalf of a particular client.

Although it would be both enlightening and interesting to study the agency in its entirety, we will define the scope of our case study by limiting it to one division within the agency. Studying one of the divisions will provide us with plenty of material for examples and problems. Later on, you can apply the skills you learn dealing with a piece of the agency to the entire enterprise.

The division we have chosen is Children's Services. And even within this division, we will examine only one part of the entire operation.

CHILDREN'S SERVICES

The Division of Children's Services provides special training, equipment, and reading material for visually impaired schoolage children. Almost all expenditures on behalf of a client are educational in nature. Education, in this sense, includes not only traditional classroom learning, but also learning about the world, learning how to move in space, learning how to organize one's materials, learning communication skills, learning social behaviors, and myriad other things.

READING MATERIAL   In terms of traditional education, though, one important topic is the availability of reading material. For a visually handicapped child, this often means a reading medium different from regular ink print. Reading material may have to be printed in a larger format (called *large print*), it might be *brailled,* or it might be *recorded* orally on cassette tape. The preferred reading medium varies with the preferences of the client.

In order to appreciate the problem Children's Services was having with school textbooks, you must first understand the differences in size, cost, and availability between a regular book and one in another medium.

Size.   A 400-page high-school textbook in regular ink print might measure $6\frac{1}{2}''\times9\frac{1}{2}''\times1''$. The same text transcribed into braille might occupy 10 to 15 spiral bound volumes each measuring $11\frac{1}{2}''\times11''\times1''$. Enlarging the text into large print at least doubles its size, and could triple or quadruple it, depending on the size of letters the client can read. Recorded on cassette tape, the book requires several tapes.

Cost.   Materials alone are costly, as are the special machines used for transcription and duplication. If the transcription is not being done by volunteers, then a single book can cost $200 to $300, and more (keep in mind that the original book cost $19.95).

Availability.   If cost were the only problem, then the obvious solution would be to spend more money. The larger problem, though, is the speed (or slowness) with which these special materials can be prepared. There is so much transcription to be done by so few people that turnaround time is often disappointingly slow. Books submitted for transcription in the spring might not be ready when the client needs them for school in the fall. And going through school without textbooks is a strain on both student and teacher.

Within Children's Services, the book problem was recognized as one needing prompt attention. The manual record-keeping system that had worked in the past was unable to handle the current volume of activity. As the concept of mainstreaming visually handicapped children into the public school system became popular, more clients were being served in nonresidential schools, so that the number of different textbooks needed by clients increased dramatically.

The Division Director saw three major problems with the way books were being ordered, inventoried and stored, and issued to clients.

1. *Book inventory.*   Due to poor record keeping, the only way to tell whether or not a particular book was in the Resource Center was physically to search the shelves. As a result, books were often reordered even when they were available, simply because they could not be found.

2. *Book ordering.*   Ordering books was a time-consuming task, filled with inaccuracies. Several copies of manually-prepared purchase orders were kept in several different places, and when books finally came in, it was nearly impossible to find the matching paperwork.

3. *Book retrieval.*   It was frequently easier to order a new copy of a braille or large-print book than it was to locate and retrieve a no-longer-needed copy from the client who had checked it out of the Resource Center. Little, if any, recycling of books was taking place. In times of fiscal austerity, an effort in this area could save the division a lot of money, freeing funds for other expenditures on behalf of more clients.

ACCOUNTING   Problems in the Division of Children's Services were not limited to books. Another problem was a financial one.

Not all funds for Children's Services come from one source. In fact, each year the state government sets aside money for the agency; the federal government also sets aside money in a special national fund that is allocated to states on a per capita basis; and the Division Director applies for (and receives) several additional government grants for special programs. Whether a client is eligible for funds from one source or another depends on certain eligibility criteria. Eligibility does not mean that the client receives money himself; rather it means that money can be spent on behalf of the client.

Multihandicapped children, for example, are eligible for funds from a special grant. Legally blind children are eligible for funds from a federal source that may not be used for visually impaired children. There is a federal grant set aside specifically for preschoolers, and another that can be used on behalf of any visually handicapped child.

Every expenditure made on behalf of a client must be charged against funds from one funding source or another. And every expenditure depletes the funding source a little bit. Clearly, the division wants the money to go as far as possible. One way to do this is to spend money from the most restrictive source for which a client is eligible. For example, if a client needs a brailled book and is eligible for both multihandicapped funds and general state funds, the division would earmark money from the multihandicapped funds for the expenditure. This leaves the less restrictive funds available for more clients.

For better management of these limited accounts, the Division Director needed more current information than was available. The director also needed more accurate data on each year's expenditures in order to write new grant proposals for the next year. Sadly, the kind of information needed (such as how much money had been spent on each client) was difficult, if not impossible, to obtain. This prompted the director to add two more problems to the list of problems in Children's Services.

4. *Individual case costs.* There was currently no way to tell how much money had been spent on behalf of an individual client.

5. *Funding-source accounting.* There was currently no way to tell how much money had been spent from each of the several funding-source accounts.

SPECIAL TRAINING Each client is assigned to a special teacher from the Division of Children's Services. The teacher provides some special training to the client in areas such as reading and writing braille, provides consultation for the child's regular classroom teacher to help the teacher adjust to the client's special needs, and helps the client order textbooks from the Resource Center as well as return them when they are no longer needed.

EQUIPMENT Visually handicapped students often need equipment to help them function in school. They may need tape recorders, typewriters, braille writers, closed-circuit televisions, and other items the average schoolage child does not require. These items can be purchased, within some limitations, using money from the various funding sources.

THE CASE STUDY

The five problems listed in the preceding section are the ones we will study. There were many more, but we limit ourselves to checking books in and out of the Resource Center, purchasing books that are not available in the Resource Center, retrieving books from clients who no longer need them, keeping track of the money spent on individual clients, and keeping track of the amount of money spent from each of the funding sources.

Let us look at what happens when a teacher requests a book on behalf of a client. As you read the description that follows, you can trace the documents in Figure 1-2. They are numbered to match the text.

A teacher sends (1) a Book Request to the Resource Center. The Resource Center checks to see if the book is available. If it is, (2) a Book Package is sent to the teacher. If the book is not available, then the Resource Center checks various catalogs to find out if the book has already been transcribed into the medium the client needs. If it has already been transcribed, the Resource Center sends (3) a Purchase Order to a book vendor, and (10) a Confirmation to the teacher. If the book has not already been transcribed, then the Resource Center gets (4) a print copy of the book and Instructions for Transcription from the teacher, sends (5) a Transcription Order to a transcription vendor, and sends (10) a Confirmation to the teacher.

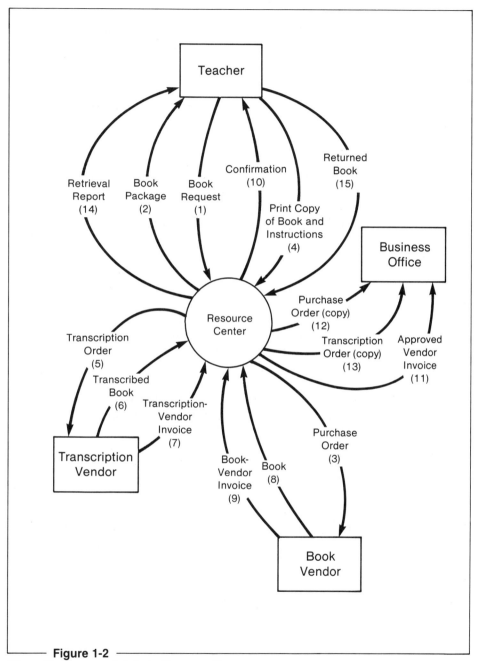

Figure 1-2

Flow of book request data in Resource Center.

The transcription vendor sends (6) the transcribed book and (7) a Transcription-Vendor Invoice to the Resource Center, or a book vendor sends (8) the book and (9) a Book-Vendor Invoice to the Resource Center.

The Resource Center sends (2) a book package to the teacher. The Resource Center sends (11) the Approved Vendor Invoice and a copy of either (12) the Purchase Order or (13) the Transcription Order to the Business Office.

Periodically, the Resource Center scans the Library Books File looking for overdue books to the teacher. It issues (14) a Retrieval Report for each overdue book. The teacher subsequently arranges for (15) the book to be returned to the Resource Center.

Each document that comes in to the Resource Center and each document that goes out has data on it. The document called a Confirmation, for example, contains the teacher's name, the client's name, the book title, the reading medium, the date on which the book was ordered, and the date the Confirmation was printed.

Likewise, a Book Request contains the teacher's name, the client's name, the date the book is needed, the title of the book, the author, the copyright date, the reading medium of the book, and a note indicating either that the client needs the book now or wants to reserve it for a future date. Figure 1-3 illustrates a shorthand way of presenting the composition of these two documents.

Ultimately the Director and the system analyst agreed on six major activities that the new system must perform (see Figure 1-4):

1. fill book requests from teachers;

2. order books—either books that are already available in the reading medium the client needs, or transcriptions of books into the medium the client needs;

3. receive books from vendors and put them into circulation;

4. issue retrieval reports for overdue books to get them back from clients;

5. put returned books back into circulation;

6. approve vendor invoices as they come in, so that the vendors can be paid.

CONFIRMATION = Teacher + Client + Title + Medium + Date Ordered + Today's Date

BOOK REQUEST = Teacher + Client + Date Needed + Title + Author + Copyright + Medium + [needed now | reserve]

Figure 1-3

Composition of Confirmation and Book Request.

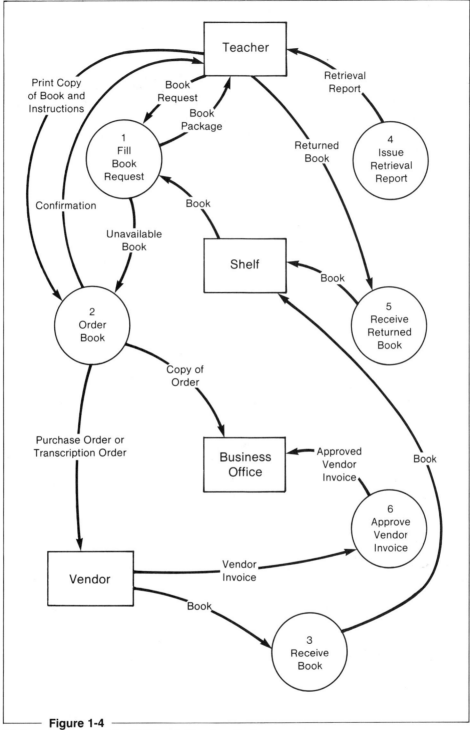

Figure 1-4

Major activities within Children's Services System.

All these activities involve access to various pieces of data related to clients, library books, available funds, and so forth. Collections of data, or files, are often accessed by many activities. As an example, consider the activities of filling a book request, issuing a retrieval report, and receiving a returned book. Each of the activities needs data in the Library Books File. Figure 1-5 is a picture of these activities with respect to the Library Books File.

When a book is checked out of the Resource Center, the client's name and the date due (1) are recorded on the Library Book Record. Periodically, the

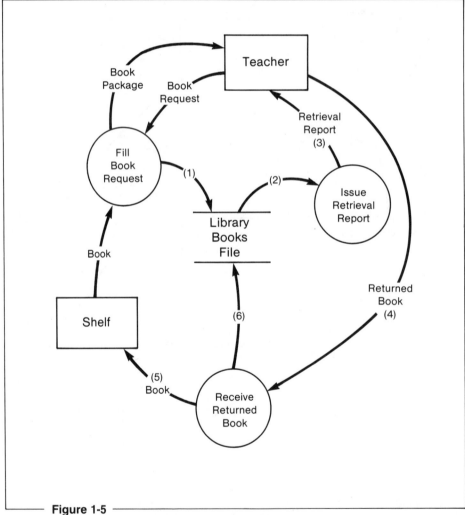

Figure 1-5
Activities using Library Books File.

Library Book File is scanned, and (2) the records for overdue books are pulled. (3) Retrieval Reports are issued and distributed to teachers, who arrange for (4) the books to be returned. Finally, the book itself (5) is placed back on the shelf, while (6) the Library Book Record in the Library Books File is updated to indicate that the book is again available.

Figure 1-5 is called a dataflow diagram, and contains four symbols. A rectangle, or *terminator,* represents someone outside the system being studied who either sends data in or receives output from the system. Circles indicate *processes.* Named arrows, called *dataflows,* show data passed from one process to another, to or from terminators, and to or from files. A pair of parallel lines represents a *file.* Notice that dataflows to and from files do not have to be named, because the name of the file implies the name of the records found in the file. In the last part of this chapter, we examine each of the symbols more closely.

There are five files in which data for the Resource Center activities are stored: Library Books, Client, Funding-Sources Accounts, Transcription Orders, and Purchase Orders. Each file consists of many records: the Client File contains Client Records, the Library Book File contains Library Book Records, and so forth. Records, like the documents illustrated in Figure 1-3, are described in terms of their contents.

In order to describe records accurately, we need some new symbols. You can see examples of each of these in Figure 1-6, which shows the contents of the Client Record and the Library Book Record.

Underlined data items are the key fields by which specific records can be accessed: Name.

Braces { } indicate that a data item might be repeated within the record. A client can check many books out of the Resource Center at one time: {Book on loan}. And the client can be eligible for funds from several funding sources: {Funding Source}.

Parentheses ( ) indicate that a data item is optional. A library book may or may not be reserved, for example, so the reservation data is optional: (Reservation).

CLIENT RECORD = Name + Date of Birth + Address + Teacher + {Reading Mode} + {Book on Loan} + {Expenditure} + {Funding Source}

LIBRARY BOOK RECORD = Title + Author + Copyright + Publisher + Catalog Number + Current Status + Current Borrower + (Reservation) + Medium + Number of Volumes + Date Out + Date Due

Figure 1-6
Composition of Client Record and Library Book Record.

Figure 1-7 shows a more complete picture of the activities in the Resource Center and their interfaces. The illustrations presented so far (Figures 1-4 and 1-5) have described only a part of the total Resource Center activities. Real-world business systems are seldom as simple as the activites in Figure 1-4. Often there are many documents or pieces of data passed from one activity to another. In addition, there are frequently data files shared by many activities. The picture in Figure 1-7 looks far more complex than anything we have seen so far. Yet it is simply an illustration of the Resource Center in terms of its processes and the data passed between them.

A typical activity is made up of a group of smaller activities—that is, the whole activity is equal to the sum of all its subactivities. We can talk about preparing a meal, for instance. Or we can talk about the separate activities of preparing the entree, vegetables, and dessert. At another level we can talk about the details involved in making dessert: preparing a batter, preparing a filling, baking them together, then mixing frosting to put on top.

So it is with business activities. Task 1, Fill Book Request, is really the name of a group of little activities that collectively fill book requests. The steps for Task 1 are expanded and illustrated in Figure 1-8. First, check to see if the book is available in the Resource Center (1.1 Determine Availability In-House). If the book is unavailable (Unavailable Book), the request cannot be filled. If it is available (Book Needed), then if the client wants to reserve the book for a future date, mark that information on the Library Book Record (Reservation Data); on the other hand, if the client needs the book right away, then check the book out of the library (1.2 Check Book Out) by recording the borrower's name on the Library Book Record (Current Borrower), and the book name on the Client Record (Book on Loan). When that's done, print a packing slip (1.3 Print Packing Slip; braille books are much larger than regular print books, and often require large cartons to hold them). Finally, pack the book (1.4 Pack Book) in boxes, attach the packing slip, and send it to the teacher.

The activities inside each of the four circles, or bubbles, in Figure 1-8 can be described in a few short sentences. Each bubble represents one small function within the overall function of filling a book request. Drawing sets of pictures that give us more or less of the details in a system is a very useful technique. What we have done is *partitioned* the system into major functions, then partitioned each of those functions into more specific functions, and partitioned those functions into smaller functions and so on, until we reached the level of detail in which a function could not be further subdivided. A function that cannot be further subdivided is instead described in detail. The bubble called Determine Availability In-House, for example, can be described using a variety of methods. Two of them are illustrated in Figures 1-9 and 1-10.

Collectively, the three documents—the pictures, the description of data compositions, and the descriptions of bubbles that are not further subdivided—describe the user's business activities. Now it is time to learn some general terminology for parts of the system specification documents.

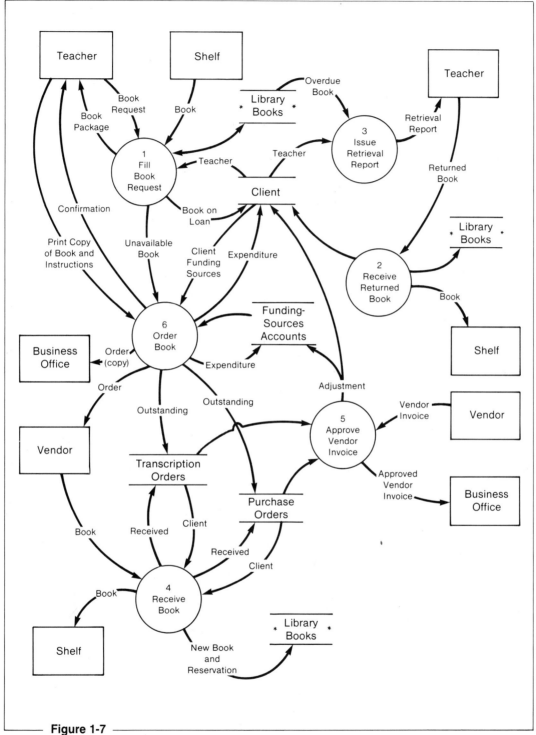

Figure 1-7
Flow of data in Resource Center.

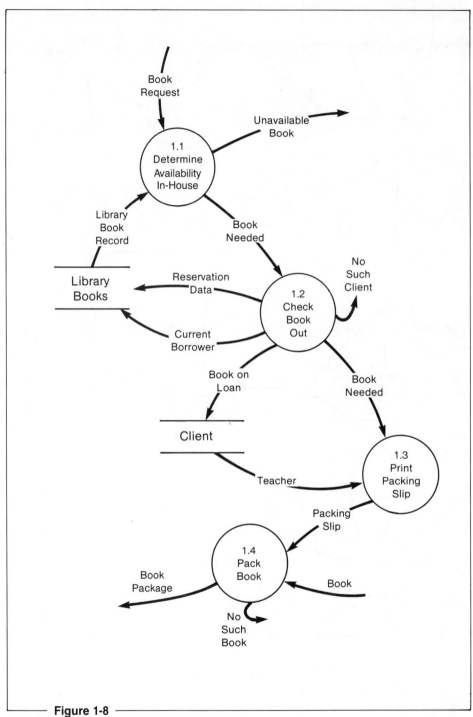

Figure 1-8

Detailed activities of FILL BOOK REQUEST.

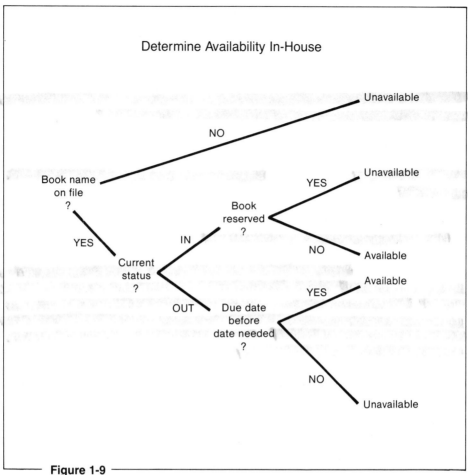

Determine Availability In-House

Unavailable

NO

Book name
on file
?

Unavailable

YES

Book
reserved
?

YES

IN

NO

Available

Current
status
?

Available

OUT

Due date
before
date needed
?

YES

Available

NO

Unavailable

Figure 1-9

Description of Determine Availability In-House (decision tree).

Determine Availability In-House
1. Using book Title
    retrieve book record from Library Books File
    IF no record for this book
    THEN book is unavailable
2. IF Current Status is In
    and book not reserved
    THEN book is Available
3. IF Current Status is Out
    and Date Due is before Date Needed
    THEN book is Available
4. OTHERWISE
    book is Unavailable

Figure 1-10

Description of Determine Availability In-House (narrative).

THE SYSTEM SPECIFICATION

The system-specification package contains three documents: a set of dataflow diagrams, a set of process specifications, and a data dictionary.

THE DATAFLOW DIAGRAM

A dataflow diagram (Figure 1-11) is a picture of data traveling through a group of processes. It is changed by the processes it goes through and sometimes it is stored for later use. We noted earlier that there are four symbols in a dataflow diagram. We now give a more complete definition of each one.

The *process bubble* represents an activity. The name of the bubble describes the activity (usually with an active verb) and the object of the activity (usually with a noun). The *dataflow* is a named arrow that represents data going into a process bubble (called *input*) or coming out of a process bubble (called *output*). The *terminator* is a rectangle representing a person, system, enterprise, or other entity that either puts into the system data to be processed or receives data the system produces. A *file*, represented by a pair of parallel lines, is used to store data until it is needed.

A system specification usually contains a set of dataflow diagrams. The first dataflow diagram, called Diagram 0, shows only the major functions or activities within the system, and how they interface. Each function is arbitrarily numbered for reference. Each function can then be *leveled* (Figure 1-12)—that is, we can draw another dataflow diagram showing the details inside one major function defined on Diagram 0. The bubbles on this lower level are numbered to match the bubble that they describe: for example, the lower-level dataflow diagram for Bubble 2 contains bubbles numbered 2.1, 2.2, 2.3, and so on. Lower-level bubbles can themselves be leveled. The terms *parent* and *child* are sometimes used to indicate the relationship between the higher and lower levels. Parent and children bubbles must *balance*—that is, all input dataflows to a parent bubble must be accounted for in its children, and all output dataflows produced by a parent must be produced by one or more of its children. A bubble that is not further subdivided is called a *functional primitive*.

THE PROCESS SPECIFICATION

A process specification is the precise description of the activities that take place inside a functional primitive bubble. It states unambiguously how the functional primitive changes its input dataflows to its output dataflows. Figures 1-9 and 1-10 are examples of process specifications. Theoretically, the entire system can be represented with one large dataflow diagram consisting only of functional primitives, each one described by a process specification. We level up (gather several children together into a parent bubble) so we can get an overview of the system without being overwhelmed with all the little details that functional primitives show us.

21

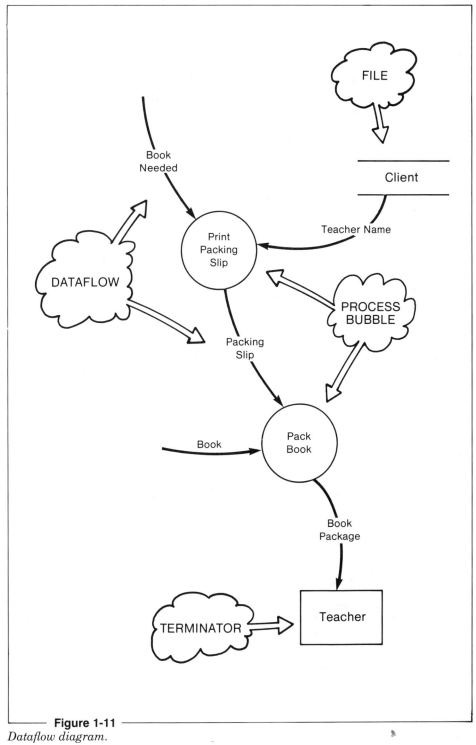

Figure 1-11
Dataflow diagram.

22

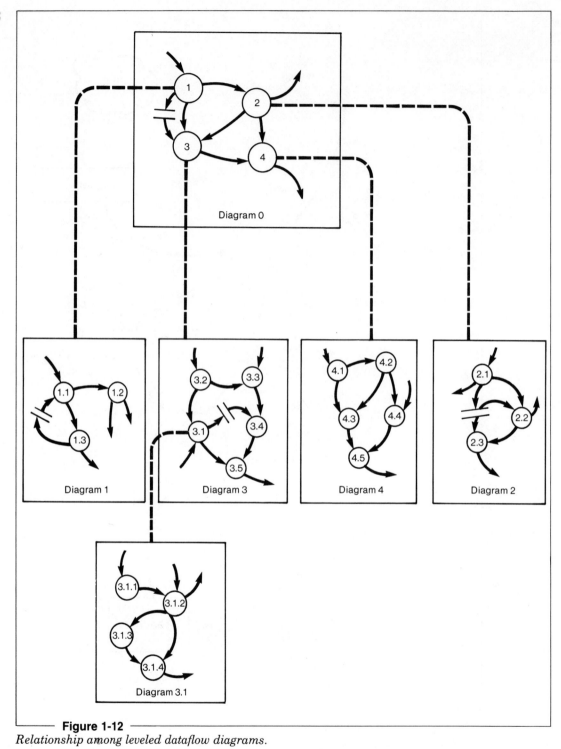

Figure 1-12

Relationship among leveled dataflow diagrams.

THE DATA DICTIONARY

The data dictionary states the composition of every dataflow and data store named on the set of leveled dataflow diagrams (Figure 1-3 and Figure 1-6). The data dictionary is usually arranged in alphabetical order so a reader can quickly locate the description of any data item.

All the pieces of the system specification fit together into a complete and cohesive description of the user's business activities (Figure 1-13). The dataflow diagram shows the data interfaces (described in the data dictionary) between processes (described in process specifications).

The appendix contains the complete system specification for the Agency for the Blind Resource Center.

SUMMARY

A computer system is made up of five components: hardware, programs, data, procedures, and people. Each of the components must be addressed when a computer system is being developed.

A system is developed in four steps: specifying the requirements of the system; proposing alternative solutions to the problem, evaluating them, and selecting one for further development; designing the system; and implementing the system. Specifying requirements identifies *what* must be done; proposing alternatives and selecting one identifies what the user feels is the *best solution* for the particular circumstances; designing defines precisely *how* the system will be built; and implementing is the *actual construction* of the system.

Specifying system requirements includes describing the user's business activities and their interfaces. The document that a system analyst and user jointly derive is called a system specification. It contains three separate but related documents: a leveled set of dataflow diagrams, a data dictionary, and a set of process specifications.

A dataflow diagram is a picture of the user's business activities and their interfaces—that is, the data passed between them. Named bubbles represent *processes*. Named arrows represent data and are called *dataflows*. Parallel lines represent stored data and are called *files*. Rectangles represent the entities that send into the system data to be processed or receive output that the system produces. They are called *terminators*.

Processes can be partitioned into collections of subprocesses in order to show more details of a single function. The bubbles that are leveled are called *parents*, while the subprocesses are called *children*. Processes that are not leveled are called *functional primitives*.

Functional primitives are described in detail by means of *process specifications*. There is a process specification for every functional primitive. Process specifications are not written for parents.

The composition of every dataflow and every file named in the set of dataflow diagrams is found in the *data dictionary*.

Together, the dataflow diagrams, the process specifications, and the data

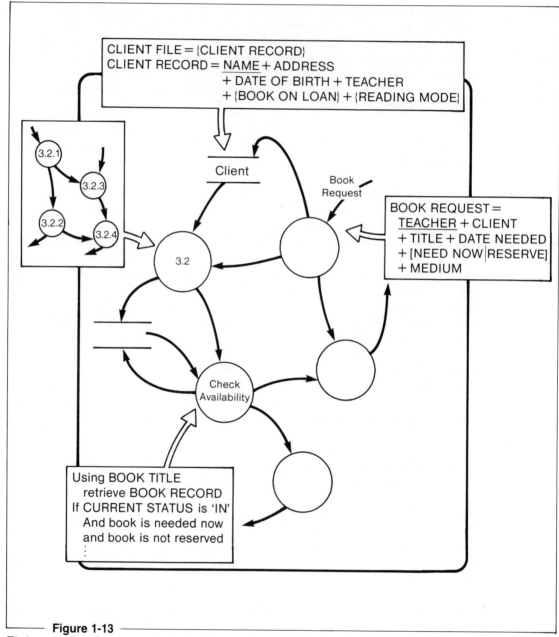

CLIENT FILE = {CLIENT RECORD}
CLIENT RECORD = <u>NAME</u> + ADDRESS
+ DATE OF BIRTH + TEACHER
+ {BOOK ON LOAN} + {READING MODE}

Client

Book Request

BOOK REQUEST =
<u>TEACHER</u> + CLIENT
+ TITLE + DATE NEEDED
+ [NEED NOW|RESERVE]
+ MEDIUM

3.2

Check Availability

Using BOOK TITLE
retrieve BOOK RECORD
If CURRENT STATUS is 'IN'
And book is needed now
and book is not reserved

Figure 1-13

Fitting together documents in the system specification.

dictionary represent a paper model of the user's business. Some of the activities will be automated in the system yet to be designed and built, and other activities will continue to be performed as they are now.

By deriving a system specification, the analyst and the user obtain a clear picture of what must ultimately be accomplished. This becomes the target for the next three steps of system development.

KEY WORDS

Analyst   See *system analyst*.

Bubble   See *process bubble*.

Child bubble   A subprocess.

Data   Facts stored in a computer.

Data dictionary   A document in the system specification. It contains an entry for each dataflow and each file, stating its name and composition.

Dataflow   An arrow in a dataflow diagram that shows data passing between processes, between a process and a file, or between a process and a terminator. Dataflows are usually named.

Dataflow diagram   A document in the system specification. It illustrates data interfaces between processes within a system.

Design   The step of the system-development process during which detailed plans for the implementation of all five system components are made.

Diagram 0   The first dataflow diagram in a set of dataflow diagrams. Diagram 0 shows the major functions of the system being modeled.

File   A collection of data represented with parallel lines on a dataflow diagram.

Functional primitive   A process bubble that is not leveled. It is described with a process specification in the system specification.

Hardware   The computer and related equipment.

Implementation   The step in the system development process during which all components of the system are built or purchased, installed, and tested.

Leveling   The activity of showing the details within a process by drawing another dataflow diagram. The process being described is called a parent and the subprocesses are called its children.

Leveling down   See *leveling*.

Leveling up   The activity of hiding details in a dataflow diagram by superimposing a process bubble (a parent) on a group of related subprocesses.

Parent bubble   A process bubble that is leveled.

Procedures   Directions that people follow in order to use a computer system.

Process bubble   A circle in a dataflow diagram that represents an activity within the system being modeled.

Process specification   A document in the system specification. It describes in detail the activities inside a functional primitive.

Program   The set of instructions a computer follows.

System analyst   An individual responsible for documenting system requirements and proposing alternative solutions. Analysts are usually skilled in data processing.

System specification   A paper model of the system being studied; consists of a leveled set of dataflow diagrams, a data dictionary, and a set of process specifications.

Terminator   An entity outside the system being modeled that either provides input or receives output; represented on a dataflow diagram with a rectangle.

User   An individual who uses a computer system.

EXERCISES

1. Name the five components of a computer system.

2. Name the four steps in the system-development process.

3. Name the three documents that make up a system specification.

4. What is the relationship between the data dictionary and the dataflow diagrams?

5. What is the relationship between the process specifications and the dataflow diagrams?

6. What is the relationship between the data dictionary and the process specifications?

7. What is the number of the dataflow diagram on which one would find the children of parent bubble 5? Of parent bubble 6.2? Of parent bubble 4.1.3? What is the number of the dataflow diagram on which one would find the parent of children bubbles 3.1, 3.2, 3.3, 3.4, and 3.5? Of children bubbles 6.4.1, 6.4.2, and 6.4.3?

8. Write the following data dictionary entries in English:

   Purchase-Order = {Line-Item}
   Insurance-Policy = [Whole-Life | Term]
   Line-Item = Number-of-Units + Catalog-Number + Description
                 (+ Size) (+ Color) + Unit-Price + Extended-Price

9. Study the specification documents in Figures 1-E9(a)–(c) and answer the following questions.
   (a) Draw the parent diagram for Figure 1-E9(a).
   (b) Why does the dataflow called Cumulative Average go in two directions when it comes out of Calculate Cumulative Average? Could the dataflow diagram be drawn any other way? What is the difference?
   (c) If the college decides to change the credits earned for letter grades (for example, if D earns 0 points and F earns −1), which document in the system specification will be changed?
   (d) If the college changes its policy toward determining honors and probation status, which documents in the system specification, if any, will be updated?
   (e) Does Format Response get enough input data flows to do its job? If not, fix it.

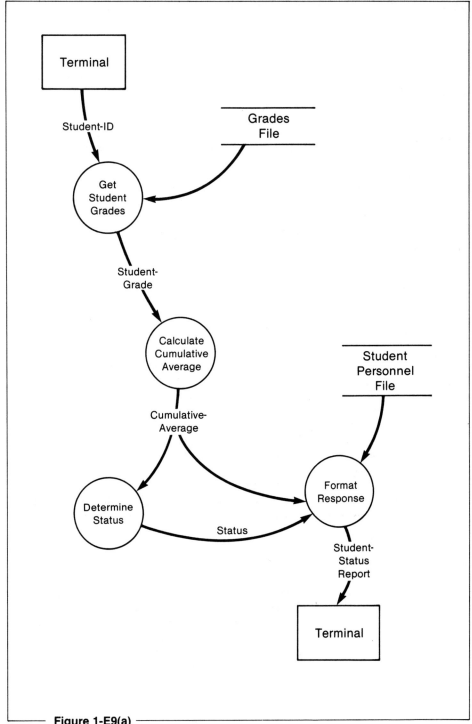

Figure 1-E9(a)

Dataflow diagram for PRODUCE STUDENT STATUS REPORT.

CUMULATIVE-AVERAGE = Numeric value indicating student's academic achievement

GRADES FILE = {Student-ID + Student-Grade}

STATUS = [blank | 'probation' | 'honors' | 'high honors']

STUDENT PERSONNEL FILE = {Student-ID + Name + Address (+ Religious-Affiliation) + Major-Field + Financial-Data}

STUDENT-STATUS REPORT = Student-ID + Name + Cumulative-Average + Status

STUDENT-GRADE = Course-Number + Credit-Hours + Letter-Grade

STUDENT-ID = Unique identification number assigned to each student

──── **Figure 1-E9(b)** ────

Data Dictionary.

Get Student Grades

Using Student-ID, retrieve grade record from Grades File.

Calculate Cumulative Average

For each grade earned by a student—

1. using the Grade Point Table below, convert letter grade to equivalent credits-earned;
2. calculate cumulative average as follows:

$$\text{Cumulative-Average} = \frac{\Sigma \, \text{Credit-Hours}}{\Sigma \, \text{Credits-Earned}}$$

3. round Cumulative Average to the nearest tenth (Example: 3.45 becomes 3.5, while 3.24 becomes 3.2).

──── **Figure 1-E9(c)** ────

Process Specifications.

GRADE POINT TABLE	
Letter Grade	Credits Earned
A	4
B+	3.5
B	3
C+	2.5
C	2
D	1
F	0

──── **Figure 1-E9(c)** *(continued)* ────

Determine Status

IF Cumulative-Average is between	THEN	the Status is
4.0 and 3.8		'high honors'
3.7 and 3.5		'honors'
3.4 and 2.0		blank
1.9 and 0		'probation'

Format Response

Using Student-ID retrieve student's Name from Student Personnel File.
Display Student-ID, Name, Cumulative-Average, and Status on terminal.

Figure 1-E9(c) *(continued)*

PART 1

Designing Programs

CHAPTER 2

The Structure Chart

When you finish this chapter, you will be able to—

- explain the function of a structure chart
- name the three graphics on a structure chart
- list the characteristics of a module
- identify boss and worker modules on a structure chart
- identify input and output parameters on a structure chart

INTRODUCTION

In this chapter we begin our study of computer-system design by examining the primary design document for the program component. As we design programs, or software, for a new system, we have in mind several short- and long-range goals. We want to—

produce programs on time and within budget;

write programs that interface correctly with each other;

make the software easy to modify, allowing for expansion as well as changes to the programs;

make it easy to trace program errors to their source and correct them.

We can achieve all these goals if we design programs properly before writing any code. Experience has shown that some programs are easier to write, understand, modify, and debug than others. Such programs usually have some, if not all, of the following characteristics:

the program solves only one specific problem;

the program contains a controlling routine in which the reader can find an overview of the program logic;

each subroutine within the program performs only one well-defined function;

the program either contains or makes reference to utility modules—standard routines used throughout the system—that are coded only once and then used over and over again;

each subroutine has access only to the specific data it needs to do its job;

the controlling routine is unaffected by minor specification changes or changes in file structure or data structure.

When we design programs we strive to incorporate as many as possible of the characteristics just listed, because doing so will enable us to produce programs quickly, get them running, debug them, and modify them in the future. The structure chart, as we will soon see, enables us to design programs that have these desirable characteristics.

In this chapter we examine structure charts and the symbols used in them. In subsequent chapters we will learn how to derive a structure chart directly from a dataflow diagram, refine it, and critique it.

A STRUCTURE-CHART EXAMPLE

Figure 2-1 shows a structure chart for a program that posts new grades to student academic records.

Each rectangle represents a routine, or *module,* in the program. A module is a group of instructions that has a single entry point and a single exit point. It has a name so other modules in the system can invoke it. It can receive input parameters when it is called, and it can return output parameters to the module that called it. It can also contain local, or private, data areas—areas to which it alone has access.

The module at the top of Figure 2-1, POST STUDENT GRADES, is the controlling routine within the program. Its name indicates the function of the entire program. The modules underneath POST STUDENT GRADES are subroutines: either internal ones coded within the program, or external ones that have been separately coded, compiled, and stored in a library for general use.

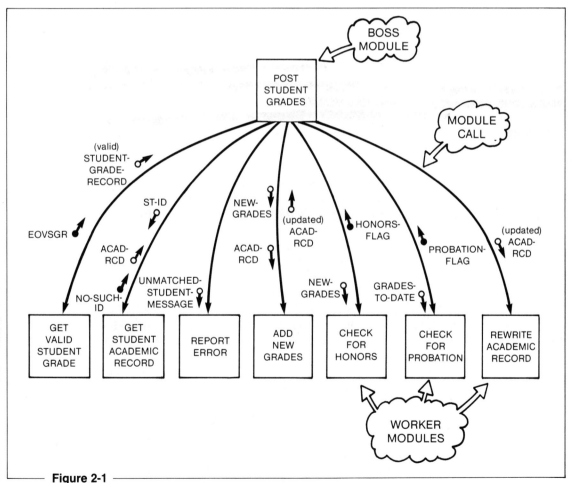

Figure 2-1

Structure chart for Post Student Grades program.

POST STUDENT GRADES is called a *boss* module because it calls other modules to help it do its work. Each of the other modules is called a subroutine, or *worker*.

Each subroutine performs a single function, indicated by its name. For example, GET STUDENT ACADEMIC RECORD gets student academic records for the boss, and CHECK FOR PROBATION checks to see if the student should be placed on probation and tells the boss.

The long arrow from the boss to each worker module indicates that the boss must *invoke* it before a worker module can do its job. Thus, the boss is responsible for executing each of the worker modules.

The small named arrows between boss and worker modules identify data and flags, called *parameters*, that bosses and workers typically pass to one another. In Figure 2-2 we have isolated two of the modules from Figure 2-1, POST STUDENT GRADES and GET STUDENT ACADEMIC RECORD. When the boss calls the worker, it passes it a student identification number, ST-ID. GET STUDENT ACADEMIC RECORD uses ST-ID to search the academic history file looking for a matching student identification number. When it finds the matching student record, it puts a copy of it into ACAD-RCD and sends it to the boss, thereby completing its job. Both ST-ID and ACAD-RCD are application data. They are therefore called *data parameters* and are illustrated on a structure chart with small hollow arrows.

The other field, NO-SUCH-ID is a *flag* and is illustrated with a small filled-in arrow. When the worker module finds a matching record it sets NO-SUCH-ID to 'N', indicating to the boss that it was successful. When the worker module cannot find a matching record it sets NO-SUCH-ID to 'Y', indicating to the boss that there is no matching academic record. NO-SUCH-ID is a flag—a field whose value indicates the status or outcome of some process. A flag in a computer program might correspond to a check in a box on a form (or even a mental note by an employee) in a nonautomated system.

The parameters going into a worker module from a boss are called the worker module's *input parameters*. Parameters going from a worker to its boss are called the worker module's *output parameters*. The words "input" and "output" do not refer to data files.

A module's name is a statement of its job. It indicates what the module does each time it is called by a boss. If the module is a boss module, then its name also summarizes the activities of all the worker modules that report to it. A module name is composed of a strong active verb and a singular specific direct object. The verb indicates precisely what the module does, and the direct object indicates what it does it to. There may also be one or more modifiers if they help clarify the module's function.

Some worker modules are themselves bosses (see Figure 2-3). Each of the lower-level modules performs a single well-defined function; its name states precisely what that function is. The boss, in this case GET VALID STUDENT GRADE, is in control of the other three modules: it calls each one when needed, passes it precisely the data it needs to do its job, and receives the results that are returned, if there are any.

37

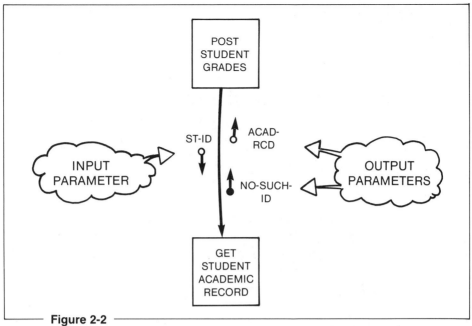

Figure 2-2
Parameters.

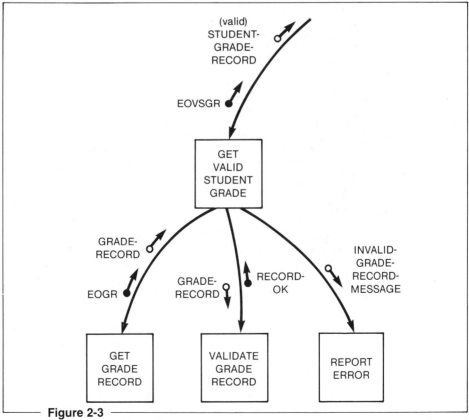

Figure 2-3
Worker module that is also a boss.

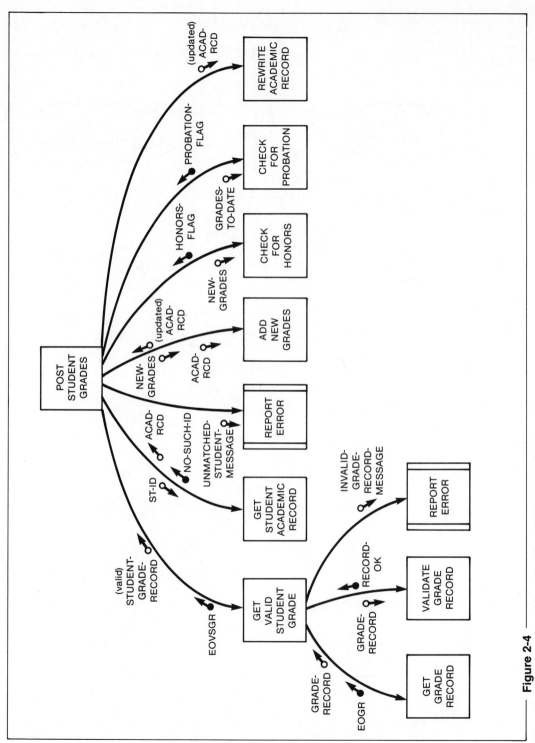

Figure 2-4

More detailed structure chart for Post Student Grades program.

Figure 2-4 shows a more detailed version of Figure 2-1. The module called REPORT ERROR is called by two bosses, POST STUDENT GRADES and GET VALID STUDENT GRADE. Modules that are called by more than one boss are called *utility* modules. They will be coded once and placed in a module library for use by any module in the system.

Another utility module is CALCULATE GPA, which appears in Figure 2-5. Like REPORT ERROR, it is called by more than one boss. And like REPORT ERROR, it is a candidate for the module library. The only restriction on a utility module is that all bosses must pass it the same number and type of parameters. On a structure chart, the names of the parameters passed by different bosses to a utility module frequently differ, because we use the names by which the bosses know the data. Vertical stripes in a module indicate that it is a library routine (see Figure 2-6).

Visualizing modules is sometimes easier if we are able to examine some actual code. We have illustrated three modules, POST STUDENT GRADES, CHECK FOR HONORS, and CHECK FOR PROBATION, and have coded them as a COBOL program and two COBOL subprograms. Even if you are not familiar with the COBOL language, you should be able to follow these very short modules (see Figures 2-7, 2-8, and 2-9).

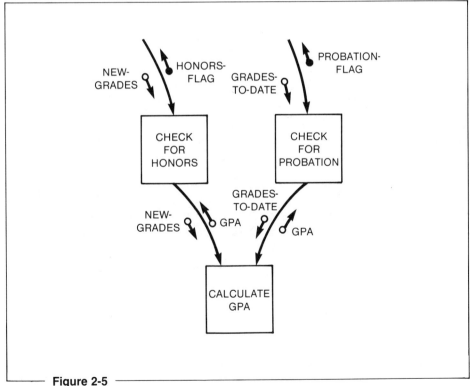

Figure 2-5
Utility module.

40

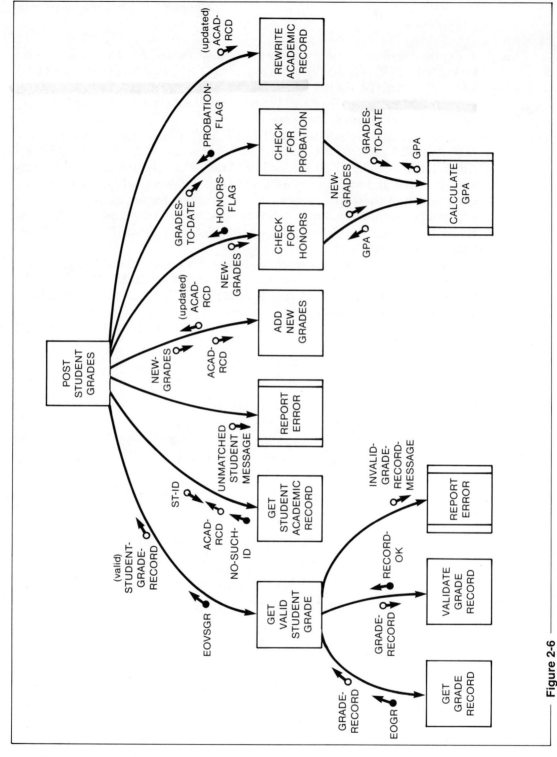

Figure 2-6
Final structure chart for Post Student Grades program.

```
IDENTIFICATION DIVISION.
PROGRAM-ID. POSTSTUDENTGRADES.
ENVIRONMENT DIVISION.
DATA DIVISION.
WORKING-STORAGE SECTION.
01   STUDENT-GRADE-REC.
05   STUDENT-ID                               PIC X(5).
05   NEW-GRADES                               PIC X(84).
01   EOVSGR                                    PIC X.
01   ACADEMIC-RECORD.  05  FILLER              PIC X(79).
                       05  PROBATION-STATUS    PIC XX.
                       05  HONOR-STATUS        PIC X.
                       05  GRADES-TO-DATE      PIC X(700).
01   NO-SUCH-ID                                PIC X.
01   UNMATCHED-STUDENT-MESSAGE.
     05  FILLER                                PIC X(15) VALUE 'NO SUCH
                                               STUDENT'.
     05  UNMATCHED-RECORD-AREA                 PIC X(89).
01   HONORS-FLAG                               PIC X.
01   PROBATION-FLAG                            PIC XX.
PROCEDURE DIVISION.
     CALL 'GETVALIDSTUDENTGRADE' USING         STUDENT-GRADE-REC
                                               EOVSGR.
     PERFORM POSTGRADE THRU POSTGRADE-EXIT UNTIL EOVSGR = 'Y'.
     GOBACK.
POSTGRADE.
     CALL 'GETSTUDENTRECORD' USING             STUDENT-ID
                                               ACADEMIC-RECORD
                                               NO-SUCH-ID.
     IF NO-SUCH-ID = 'Y'
         MOVE STUDENT-GRADE-REC TO UNMATCHED-RECORD-AREA
         CALL 'REPORTERROR' USING              UNMATCHED-STUDENT-MESSAGE
         CALL 'GETVALIDSTUDENTGRADE' USING STUDENT-GRADE-REC
                                               EOVSGR
         GO TO POSTGRADE-EXIT.
*    MATCHING RECORD WAS FOUND
     CALL 'ADDNEWGRADES' USING                 NEW-GRADES
                                               ACADEMIC-RECORD.
     CALL 'CHECKFORHONORS' USING               NEW-GRADES
                                               HONORS-FLAG.
     IF HONORS-FLAG = SPACES
         CALL 'CHECKFORPROBATION' USING        GRADES-TO-DATE
                                               PROBATION-FLAG
         MOVE PROBATION-FLAG TO PROBATION-STATUS
     ELSE
         MOVE SPACE TO PROBATION-STATUS
         MOVE HONORS-FLAG TO HONOR-STATUS.
     CALL 'REWRITEACADEMICRECORD' USING        ACADEMIC-RECORD
     CALL 'GETVALIDSTUDENTGRADE' USING         STUDENT-GRADE-REC
                                               EOVSGR.
POSTGRADE-EXIT.
     EXIT.
```

Figure 2-7

COBOL code for Post Student Grades module.

```
IDENTIFICATION DIVISION.
PROGRAM-ID. CHECK FOR HONORS.
ENVIRONMENT DIVISION.
DATA DIVISION.
WORKING-STORAGE SECTION.
01   HIGH-HONORS                      PIC S9V99    VALUE IS +3.80.
01   HONORS                           PIC S9V99    VALUE IS +3.50.
01   COMMENDATION                     PIC S9V99    VALUE IS +3.35.
01   HONORS-FLAGS.
     03   HIGH-HONORS-CODE            PIC X        VALUE IS 'H'.
     03   HONORS-CODE                 PIC X        VALUE IS 'O'.
     03   COMMENDATION-CODE           PIC X        VALUE IS 'C'.
01   MAXIMUM-GRADES-ALLOWED           PIC 99       VALUE IS 12.
01   GPA                              PIC S9V999.

LINKAGE SECTION.
01   NEW-GRADES-TABLE.
     03   GRADE OCCURS MAXIMUM-GRADES-ALLOWED TIMES PIC X(7).
01   HONORS-FLAG                      PIC X.

PROCEDURE DIVISION USING GRADES-TABLE, HONORS-FLAG.

     CALL CALCULATE-GPA USING NEW-GRADES-TABLE, GPA.

     IF GPA IS LESS THAN COMMENDATION
        MOVE SPACES TO HONORS-FLAG
     ELSE
     IF GPA IS LESS THAN HONORS
        MOVE COMMENDATION-CODE TO HONORS-FLAG
     ELSE
     IF GPA IS LESS THAN HIGH-HONORS
        MOVE HONORS-CODE TO HONORS-FLAG
     ELSE
        MOVE HIGH-HONORS-CODE TO HONORS-FLAG.

     GOBACK.
```

Figure 2-8

COBOL Code for Check For Honors module.

```
IDENTIFICATION DIVISION.
PROGRAM-ID. CHECK FOR PROBATION.
ENVIRONMENT DIVISION.
DATA DIVISION.
WORKING-STORAGE SECTION.
01   WARNING-LIMIT                  PIC S9V99   VALUE IS +1.75.
01   PROBATION-LIMIT                PIC S9V99   VALUE IS +1.50.
01   SEVERITY-LEVEL-FLAGS.
     03  SEVERE-PROBATION-CODE      PIC XX      VALUE IS 'SP'.
     03  WARNING-PROBATION-CODE     PIC XX      VALUE IS 'WP'.
     03  NO-PROBATION-CODE          PIC XX      VALUE IS 'NP'.
01   MAXIMUM-GRADES-ALLOWED         PIC 999     VALUE IS 100.
01   GRADE-POINT-AVERAGE            PIC S9V999.

LINKAGE SECTION.
01   GRADES-TABLE.
     03  GRADE-ENTRY OCCURS MAXIMUM-GRADES-ALLOWED TIMES PIC X(7).
01   RETURN-FLAG                    PIC XX.

PROCEDURE DIVISION USING GRADES-TABLE, RETURN-FLAG.

    CALL CALCULATE-GPA USING GRADES-TABLE, GRADE-POINT-AVERAGE.

    IF GRADE-POINT-AVERAGE IS LESS THAN PROBATION-LIMIT
        MOVE SEVERE-PROBATION-CODE TO RETURN-FLAG
    ELSE
    IF GRADE-POINT-AVERAGE IS LESS THAN WARNING-LIMIT
        MOVE WARNING-PROBATION-CODE TO RETURN-FLAG
    ELSE
        MOVE NO-PROBATION-CODE TO RETURN-FLAG.

    GOBACK.
```

Figure 2-9
COBOL code for Check for Probation module.

STRUCTURE-CHART BENEFITS

At the beginning of this chapter, we noted that programs that are easy to develop, test, debug, and modify usually have certain characteristics. Let us look now at how structure charts can help us design programs that have those characteristics.

PROGRAM SOLVES ONLY ONE PROBLEM

The name of the controlling module, the one at the top of a structure chart, indicates the overall function of the program. Because each of the modules reporting to the boss must contribute somehow to the boss's function, we can be sure the program does only one thing.

THERE IS AN OVERVIEW OF PROGRAM LOGIC

Once again we turn to the boss module. It is the controlling module at the top that calls each of the program's subroutines to do the work. Therefore the boss module contains an overview of the program logic. Although worker modules can themselves be bosses that call other lower-level modules, each worker contains an overview only of its own subfunction. Overall program logic is found only at the highest level.

CONTROLLING ROUTINE RESISTANT TO CHANGE

Because lower-level modules interface with files, format reports, perform calculations, and so forth, they are the ones that are most likely to change. The controlling module is protected from most minor specification changes as well as changes concerning files and data structure. Therefore maintenance becomes very localized. This means that the program can be easily and quickly modified without affecting major program logic found in the boss module.

EACH SUBROUTINE PERFORMS ONE FUNCTION

The name of each subroutine states its function. Any module that calls workers must call only those that perform part of its overall task—that is, a boss can call only modules that do something the boss would otherwise have to do for itself. Because each subroutine does only one specific task, understanding and maintaining the program is relatively easy.

EACH SUBROUTINE GETS ONLY NECESSARY DATA

The parameters on a structure chart indicate the specific data shared by boss and worker modules. By using structure charts, we can verify that modules receive *only* the data they really need in order to perform. Modules can have their own local data, of course. But by controlling access to shared data, we are able to protect the data from unauthorized or inadvertent use or alteration. The structure chart indicates all data that is exposed to more than one module.

UTILITY MODULES

Structure charts enable us to identify new utility modules, and to incorporate ones that already exist into the new program. Use of utility modules speeds up the development process because it eliminates duplicated coding, testing, and debugging; it also results in faster and easier maintenance by localizing changes.

SUMMARY

A structure chart is a picture of a program. It shows the overall function of the program, as well as details of each of the program's subfunctions. A structure chart contains these symbols:

   a rectangle represents a module—a named group of instructions that performs a single function each time it is called;

   long arrows connecting modules represent subroutine calls;

   small hollow arrows represent data items shared by a boss module and a worker module;

   small filled-in arrows represent flags—special fields set to certain values by modules to report on a situation that has occurred.

Use of structure charts will enable us to design programs that are easy to develop, test, debug, and maintain. The resulting programs are made up of single-function worker modules that report to a boss. The boss module contains the basic program logic and serves as an overview to the program. Subroutines have access only to the data they need, thus protecting data from unnecessary exposure. Lower-level modules that perform only minor functions will probably bear the brunt of most specification changes and file or data changes; the boss is shielded from them. On the other hand, incorporating changes is relatively easy because workers perform very specific functions; this results in localized maintenance. Finally, the use of utility modules speeds up development and eases future maintenance.

KEY WORDS

Boss module   A calling module on a structure chart.

Controlling module   The module at the top of a structure chart.

Data parameter Application data passed between two modules. It is represented by a hollow arrow on a structure chart.

Flag A special field whose value indicates some condition. It is represented by a filled-in arrow on a structure chart.

Input parameter A parameter passed from a boss module to a worker module.

Invoke To call. This is represented on a structure chart by an arrow connecting modules. A boss module invokes a worker module.

Module A named group of instructions, also called a subroutine. A module is represented by a rectangle on a structure chart.

Output parameter A parameter passed from a worker to its boss.

Parameter Data or a flag passed between boss and worker module. See *data parameter, flag*.

Software Computer programs.

Structure chart A diagram showing modules within a program, their organization, and parameters passed between them.

Subroutine See *module*.

Utility module A module that is called by more than one boss. A utility module can be coded and tested once, then placed on a library so it can be accessed by any modules that need it.

QUESTIONS

1. A structure chart is a document used for—
   (a) formatting reports.
   (b) building files.
   (c) designing a program.
   (d) verifying dataflow diagrams.

2. Each box on a structure chart represents—
   (a) a module.
   (b) an operating system.
   (c) a call statement.
   (d) a COBOL instruction.

3. The small arrows on a structure chart represent—
   (a) file data.
   (b) call statements.
   (c) functionally decomposed modules.
   (d) passed parameters.

4. A module—
   (a) shares data with its boss.
   (b) performs a single function each time it is called.
   (c) is a named group of instructions.
   (d) is all of the above.

EXERCISES

Study the structure chart in Figure 2-E5 and answer the following questions.

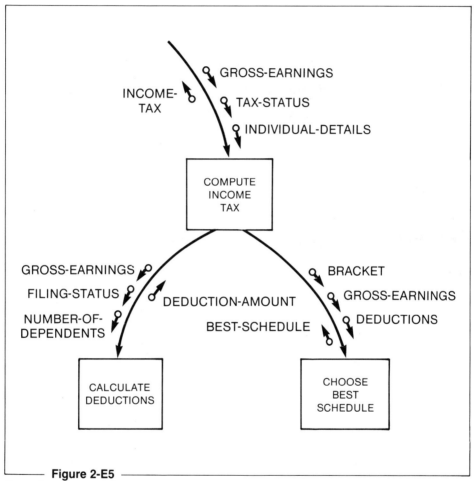

Figure 2-E5
Part of a structure chart.

5. How many parameters are passed between COMPUTE INCOME TAX and its boss?

6. What is the minimum number of call or perform statements inside COMPUTE INCOME TAX?

7. What is the name of COMPUTE INCOME TAX's boss?

8. How many parameters are passed between COMPUTE INCOME TAX and CALCULATE DEDUCTIONS?

9. Name all the parameters on the structure chart.

10. Name all the modules on the structure chart.

11. What is the output of CALCULATE DEDUCTIONS?

12. Name the input parameter(s) to CALCULATE DEDUCTIONS.

13. Is COMPUTE INCOME TAX a boss module or a worker module?

14. Is CHOOSE BEST SCHEDULE a boss module or a worker module?

15. When the Internal Revenue Service changes the policy for computing deductions, which module(s) on the structure chart will have to be changed?

CHAPTER 3

Deriving the First-Cut Structure Chart

When you finish this chapter, you will be able to—

- derive a first-cut structure chart from a dataflow diagram using transform analysis
- derive a first-cut structure chart from a dataflow diagram using transaction analysis

INTRODUCTION

A dataflow diagram is a picture of a system. It is built by tracing *known* input data through one or several processes that gradually transform the data into *desired* output. In other words, a dataflow diagram deals with real, tangible, verifiable inputs and outputs. A dataflow diagram can easily be checked for completeness and accuracy and is, therefore, an excellent tool for organizing, documenting, and refining one's knowledge of the steps involved in a process. It is also useful for defining the many subfunctions within a larger process because a dataflow diagram can be leveled. When we level, we select one bubble on the dataflow diagram, and draw another dataflow diagram illustrating the various subfunctions of that bubble.

Thus a dataflow diagram is a very useful tool for someone learning about a process, particularly a complex one, because it ensures that the learner is dealing—

with a complete set of known inputs and outputs,

with a complete set of subfunctions, and

only with subfunctions that relate to the process being studied.

When we design and subsequently write programs, we need exactly the same assurances: that we know what data are needed in order to produce the data that are expected; that we write a complete set of instructions for the computer, omitting nothing; and that we include in the program only those routines that contribute to the overall function of the program—that is, only ones related to the solution of the problem the program is supposed to solve.

It would seem logical and desirable, then, to first draw a dataflow diagram for a program, and then to write code from the diagram. There are some obstacles, however, that prevent a *direct* transition from dataflow diagram to code. First, a dataflow diagram illustrates multiple processes executed concurrently. Their only connection is the data produced by one process and passed to another. Other than that, the processes are independent of one another. Conversely, a program is made up of many groups of instructions, or modules, that pass not only data but also *control* to one another. This is true of programs because only one instruction at a time can be in control of the computer's resources. Control must be passed from one module to another like a baton in a relay race.

Second, dataflow diagrams concentrate on normal processing and virtually ignore exception (or error) processing. Programs, on the other hand, can contain many exception-handling routines in addition to routines for normal processing. Third, dataflow diagrams assume that there is a never-ending supply of input data to be processed. This assumption is quickly abandoned at the program level, because programs typically get their input data from files that contain a finite number of records. Programs, then, must account for activity that is to take place when the end of the input stream has been encountered.

These three things, passing control from one module to another, processing errors, and handling end-of-file conditions, must be coded in a program but are not found in a dataflow diagram. The structure chart, as we saw in the previous chapter, allows us to illustrate all three. In this chapter, we learn how to derive a structure chart directly from a dataflow diagram.

Deriving a structure chart this way allows us to take advantage of the characteristics of both the dataflow diagram and the structure chart. The dataflow diagram assures that we are dealing with the right inputs, processes, and outputs, and that we have neither introduced extraneous data or processes nor omitted any that are needed. The structure chart, because it is derived from the dataflow diagram, will also be complete and correct. And on it we can also illustrate control, error handling, and end-of-file conditions, which are important aspects of programming.

IDENTIFYING PROGRAM BOUNDARIES

Perhaps the most important step in this procedure is deciding which part of a dataflow diagram constitutes a program. A program is a group of computer instructions that is loaded by the operating system all at once and then executed.

Theoretically, all the functional primitive bubbles (the ones that are not leveled any further) on the dataflow diagram could be implemented directly. In other words, we could write one giant (!) program for the whole system. But that one program would be extremely difficult to manage—just reading and maintaining it could become a horrendous task. The slightest change in requirements or equipment would mean that the entire system would have to be drydocked for repairs. In some cases, there might not even be enough computer memory to contain the whole program at once (consider personal computers). Finally, if some part of this giant program malfunctioned, it could be both time consuming and costly to trace and repair the malfunctioning code. Thus, writing one giant program for the whole system is a ludicrous approach.

On the other hand we could, theoretically, write one program for each functional primitive, implement it on its own computer, and interface the computers so they could pass data to one another. This approach would eliminate some of the problems we face with the "one giant program" approach: maintenance is very localized, and computer memory is no longer a constraint. However, it could be expensive to install one computer for each and every functional primitive bubble!

Fortunately, there is a middle ground between writing one giant program and installing a separate computer for every functional primitive: break up the dataflow diagram into subsets, and then implement each subset as a separate program. The next question, of course, is "What are the best subsets?" In order to select the best subsets, we consider these factors:

online vs batch application requirements,

commercial software packages,

audit and control requirements,

frequency of use,

hardware boundaries.

ONLINE VS BATCH APPLICATION REQUIREMENTS

Some systems require some functions to be *online*—that is, the functions must be available for execution whenever they are requested. An example of an online application is an interactive airline-reservation system. Various functions—such as making, changing, and canceling flight reservations, checking flight availability, and printing tickets—must be available for execution all the time. Therefore all the functions must be loaded into the computer together, perhaps as a main program that calls a different subprogram for each function. (Even if the subprograms are not all loaded into computer memory together, they must at least be available for instant retrieval from an online storage device when they are called for.)

In contrast to this online processing, other functions can be executed in *batch* mode. Batch mode means that data are collected and processed in *groups*. Batch processing is often used when there is no need to process each input dataflow as soon as it is available, when there is a time delay between input and processing, and when data must be sorted or merged or both before they are processed. An example of batch processing is a payroll application: time cards are collected and validated in a group; the group is sorted into employee-number sequence; the employee payroll file is updated; and payroll checks and stubs are printed in a group.

COMMERCIAL SOFTWARE PACKAGES

There is an abundance of commercial software available today: accounting packages, file-management packages, graphic-display packages, library-research packages, student-records packages, and so forth. More and more companies are taking advantage of commercial software by choosing to buy rather than develop their own application software in-house.

After specifying the system requirements, we can seek commercial software that will satisfy some or all of them. Often a company discovers that there is not quite a perfect fit between the system requirements and the features of a software package. Sometimes the package can be tailored to match the system requirements; at other times the user is willing to adjust requirements in order to take advantage of a commercial package. In either case, it is usually faster and less expensive in the long run to buy software rather than to develop it from scratch.

If commercial software is found that performs some of the system functions, those functions can be eliminated from the dataflow diagram. What remains will be developed in-house.

AUDIT AND CONTROL REQUIREMENTS

Auditors often impose controls on the system being developed. They do this because they must be able to trace transactions through the system: as transactions enter, as they are posted to accounts, as they appear on output reports and statements, as they trigger other events to occur, and so forth. The auditors also need to be able to trace items backward through the system in order to determine the origin of any piece of data. Their job is to ensure that only processable data are processed, that invalid data are rejected, and that errors can be traced and reversed.

One thing auditors may require is that special reports be generated between certain processes, so that the output data from one process can be verified before the next process may be executed. The result is that what could have been accomplished in a single program may have to be subdivided into several programs, run one after the other. Because programs pass data through files, we may be forced to introduce files into the dataflow diagram that are not part of the user's requirements, but that become part of the system nonetheless. The same is true of the intermediate audit reports: they were not required by the user, but become system output documents.

FREQUENCY OF USE

Another consideration when establishing program boundaries is the frequency with which functions are executed. We already determined that some functions might be needed online and will become one program made up of several subprograms. Of the remaining functions, some will be executed with regularity (daily, weekly, quarterly, annually); others will be executed only as they are needed.

Functions that are executed at different times become separate programs. For example, the weekly payroll functions become one program, while the functions that collectively print W-2 forms at the end of the year are implemented in a separate program.

HARDWARE BOUNDARIES

Some systems incorporate a number of processors: intelligent terminals that can be programmed, microcomputers or minicomputers (front-end processors) that communicate with a host computer, or multiple computers that are not directly interfaced, but that share data files. In such systems, we may be able to distribute various functions among the processors.

For example, intelligent terminals and front-end processors might be used to capture, validate, and edit data before releasing them to the host computer for processing. Multiple computers might be used to capture and process daily transactions against their own files, then produce an activity file on tape or diskette that is shipped to a company's main computer for processing against central files.

Furthermore, certain operations can be implemented only on particular machines, such as analog-to-digital converters.

Identifying hardware boundaries is another aspect of isolating programs, because hardware boundaries separate programs. Intermediate files are not needed between machines if the machines are directly interfaced, but otherwise they are needed.

To summarize, a program is a set of instructions that is loaded into the computer all at once and executed. We select program boundaries by separating the system into online and batch applications, by purchasing commercial software to implement some system functions, by introducing controls, by studying the frequency with which functions are executed, and by identifying hardware boundaries.

We must design and specify *each program* before writing any code. Program specifications are based on a structure chart that is derived directly from a dataflow diagram. There are two major strategies we use to derive a structure chart, depending on the nature of the dataflow diagram. *Transform analysis* applies to dataflow diagrams that contain linear, or sequential, tasks. *Transaction analysis* applies to dataflow diagrams containing case-structured tasks.

TRANSFORM ANALYSIS

When we perform transform analysis, we execute the following steps.

Step 1. Select a dataflow diagram for one program.

Step 2. Identify the area of central transform.

Step 3. On another sheet of paper, rearrange the bubbles so the one identified as the area of central transform (otherwise known as "the boss") is at the top and center of the page, and then copy the rest of the dataflow diagram as if the other bubbles were dangling underneath the boss.

Step 4. Change the graphics so the diagram begins to take the shape of a structure chart: bubbles become rectangles, dataflow arrows become calls, and data-parameter arrows indicate the flow of data between modules.

Step 5. Add read and write modules to access files.

Step 6. Adjust module names in keeping with naming conventions for structure charts: the name should state *what* it does for its boss, not how it does it; add necessary flags.

In order to illustrate transform analysis, we return to the Agency-for-the-Blind case study.

In Step 1, we have selected Diagram 1, Fill Book Request, from the appendix as our example. It is reproduced in Figure 3-1.

In Step 2, we identify the *area of central transform*. We can take one of two approaches.

Approach 1. Pick the bubble that has more dataflows than any of the others.

Approach 2. Trace input streams into the dataflow diagram asking "Is it still input?" When the answer is "No," you have identified the end of the input stream. Trace output dataflows backward into the dataflow diagram asking "Is this considered output?" When the answer is "No," you have identified the beginning of the output stream. The bubble or bubbles between the end of the input stream and the beginning of the output stream comprise the area of central transform. (Intuitive, isn't it?)

We will use Approach 1 first and complete the transform analysis of this dataflow diagram. Then we will look at another example using Approach 2.

The bubble called INQUIRE AVAILABLITY IN-HOUSE deals with six dataflows. No other bubble on the page has as many inputs and outputs; INQUIRE AVAILABILITY IN-HOUSE appears to be right in the heart of the process of filling a book request. Having identified that bubble as the area of central transform, we continue with Step 3, redrawing the entire dataflow diagram with the area of central transform bubble at the top.

Figure 3-2 shows the results of Step 3. The central transform bubble is now at the top of the page, and the rest of the dataflow diagram has been copied underneath it. Nothing but the arrangement of the bubbles has been changed from the original dataflow diagram in Figure 3-1.

In Step 4, we introduce structure-chart graphics. The picture in Figure 3-3 is beginning to resemble a structure chart. We can see modules, module calls, and passed parameters.

The results of Step 5 (add read and write modules for file access) are illustrated in Figure 3-4. The parallel lines that represented the LIBRARY and CLIENT files are gone now, having been replaced with read and update modules. We have added other input and output modules as well: to get book requests from the terminal, to print packing slips, and to display messages on the terminal.

Finally, Step 6 calls for us to change module names and add flags. Figure 3-5 shows the completed first-cut structure chart for Fill Book Request. It is not yet a finished structure chart. In the next chapter, we will learn how to refine the first-cut structure chart.

56

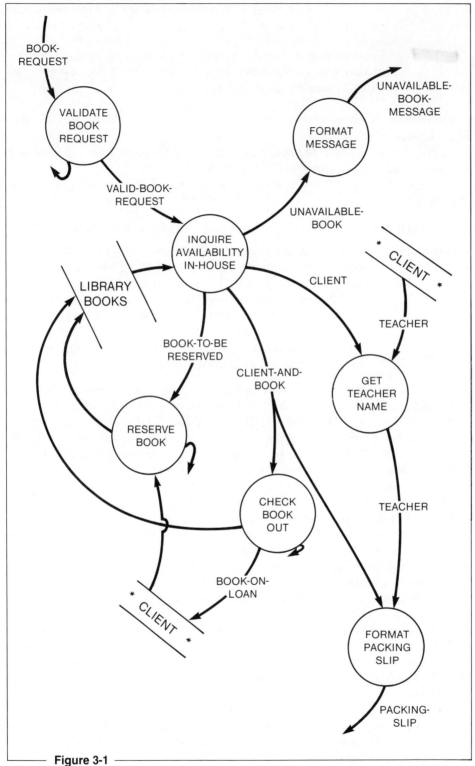

Figure 3-1

Step 1: Select dataflow diagram.

Let us walk through another example of transform analysis, this time using Approach 2 for identifying the area of central transform. Figure 3-6 shows the dataflow diagram for a program that posts new grades to a student's academic record, checks to see if the student has earned honors (in which case the parents

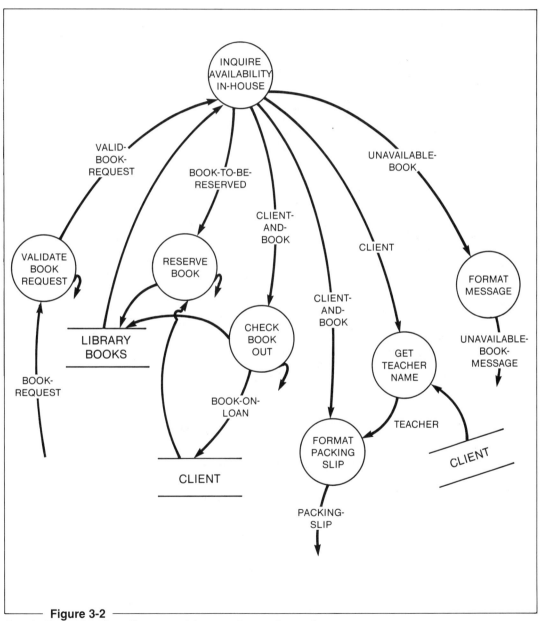

Figure 3-2

Step 3: Draw dataflow diagram with area of central transform at top.

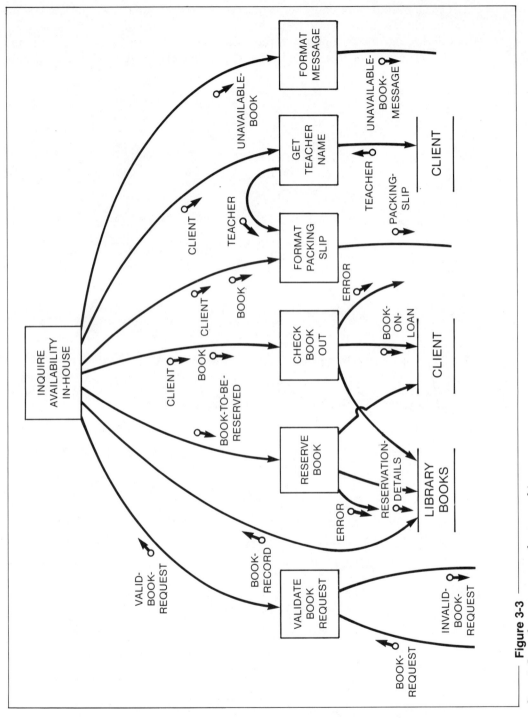

Figure 3-3

Step 4: Introduce structure chart graphics.

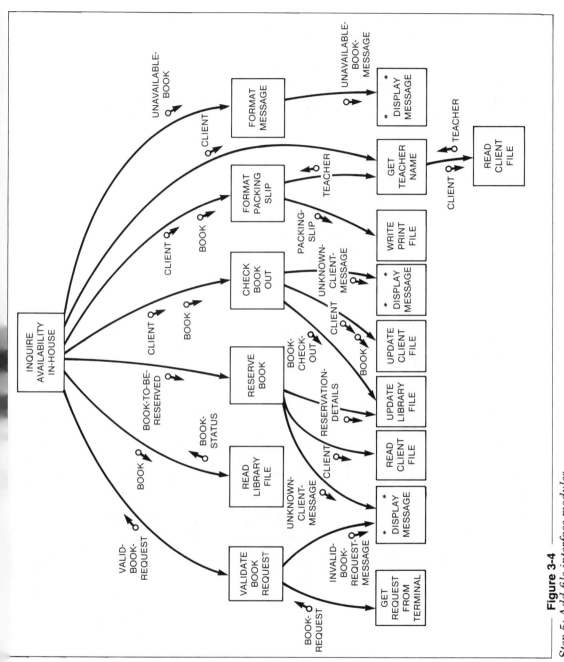

Figure 3-4

Step 5: Add file interface modules.

receive a letter of congratulations), and checks to see if the student ought to be placed on probation (in which case the student receives a probation notice). Figure 3-6 completes Step 1 (select a dataflow diagram).

Now we perform Step 2 (identify the area of central transform), using Approach 2. Starting with the input dataflow, NEW-GRADE, we follow it into the dataflow diagram asking as it passes through a bubble, "Is this still input data?" NEW-GRADE is certainly raw input data. By contrast, VALID-NEW-GRADE

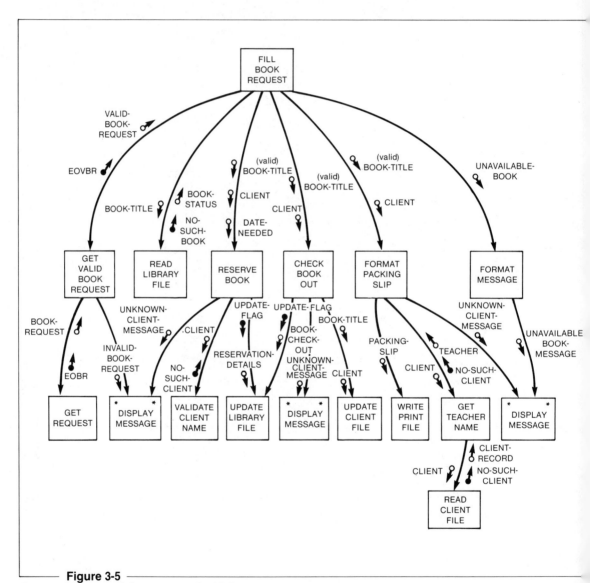

Figure 3-5

Step 6: Adjust module names and add flags.

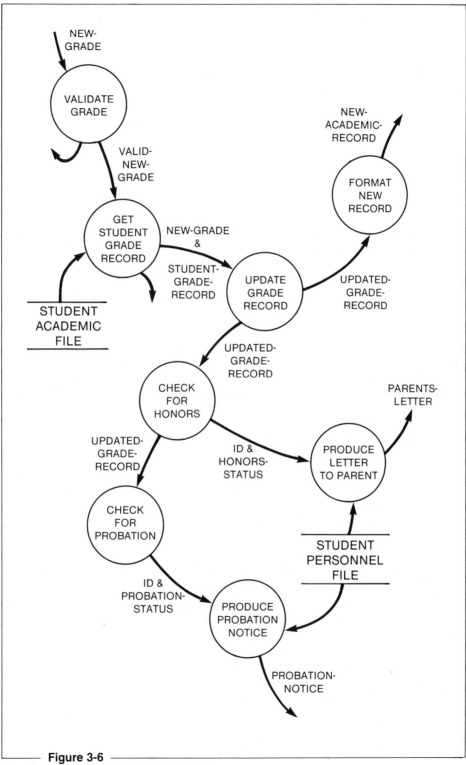

Figure 3-6

Dataflow diagram for POST GRADES.

is more logical than NEW-GRADE (it has been cleaned up a bit, and so is less physical than NEW-GRADE), but even VALID-NEW-GRADE cannot be processed until it is matched with a STUDENT-GRADE-RECORD. It appears that the end of the input stream is the dataflow called NEW-GRADE & STUDENT-GRADE-RECORD. Once we have both of these data items, we can begin the process of posting grades.

There are three output dataflows: PARENTS-LETTER, PROBATION-NOTICE, and NEW-ACADEMIC-RECORD. The point at which all information needed to produce PARENTS-LETTER is available is between CHECK FOR HONORS and PRODUCE LETTER TO PARENT. That point marks the beginning of the PARENTS-LETTER output stream. Similarly, the point at which the data needed to produce PROBATION-NOTICE is available is between CHECK FOR PROBATION and PRODUCE PROBATION NOTICE; this marks the beginning of the second output stream. Finally, the beginning of the NEW-ACADEMIC-RECORD output stream is located between UPDATE GRADE RECORD and FORMAT NEW RECORD. In Figure 3-7(a), we have identified the area of central transform as consisting of UPDATE GRADE RECORD, CHECK FOR HONORS, and CHECK FOR PROBATION.

Just as dataflow diagrams can be leveled down to show more detail, they can be leveled up to conceal detail. In Figure 3-7(b), we have collected the central transform bubbles together into a parent called AREA OF CENTRAL TRANS-FORM. This new bubble has one input dataflow, NEW-GRADE & STUDENT-GRADE-RECORD. It has three output dataflows: UPDATED-GRADE-REC-ORD, ID & HONORS-STATUS, and ID & PROBATION-STATUS.

Having identified the area of central transform, and thus the boss bubble, we can continue with Step 3 of transform analysis. This time, however, we introduce a twist: we know that the area of central transform is made up of several processes (we leveled up to hide them, but they are still there). Therefore, we know that eventually we will have to factor those functions out of the controlling module at the top of the resulting structure chart. If this is true, then why include them in the top module at all? Why not *create* a controlling module and place *it* at the top of the evolving structure chart?

Examine Figure 3-8 carefully. Notice that this time the redrawn dataflow diagram does not look identical to the original one. There are two differences: the processes within the area of central transform have been collected into one new bubble, called AREA OF CENTRAL TRANSFORM, and a controlling module called POST GRADES has been introduced at the top of the structure chart.

In Step 4, we introduce structure-chart graphics: rectangles, data-parameter symbols, and module-call arrows. In this step, we also factor the central transform into its original three processes, illustrating each one as a separate module. The diagram in Figure 3-9 is starting to look like a structure chart.

Figures 3-10 and 3-11 show the structure chart after Steps 5 and 6 have been performed. They are the same as they were in the previous transform-analysis example for filling a book request in the Resource Center at the Agency for the Blind.

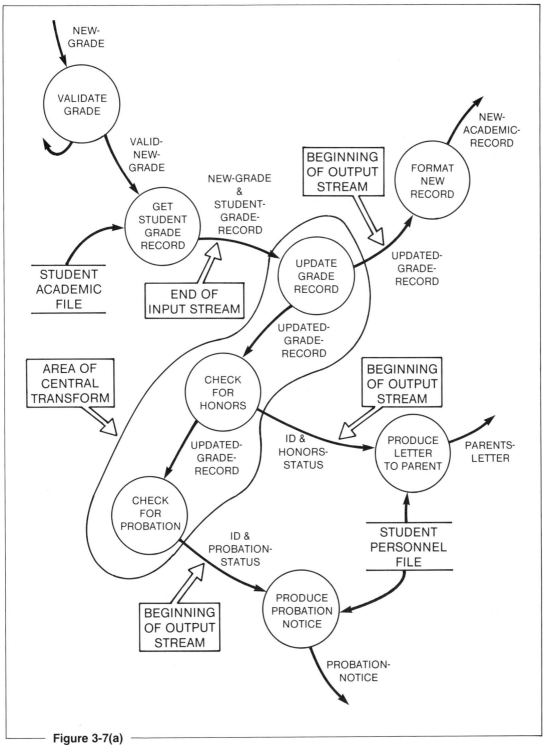

Figure 3-7(a)
Area of central transform.

64

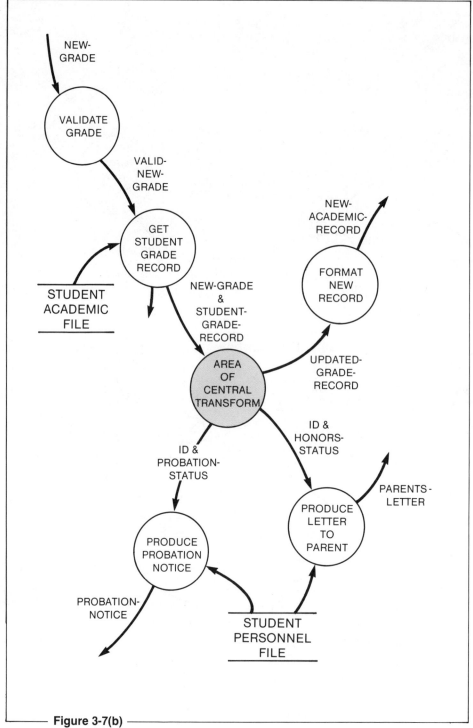

Figure 3-7(b)

Area of central transform leveled up.

Finally, Figure 3-12 illustrates the first-cut structure chart for the POST GRADES program. Although it is not necessarily a step in transform analysis (we could have stopped with Figure 3-11), in Figure 3-12 we have eliminated what we felt was an unnecessary module (AREA OF CENTRAL TRANS-FORM), thus changing the structure chart so that each of the modules in the area of central transform communicates directly with POST GRADES.

TRANSACTION ANALYSIS

Some dataflow diagrams illustrate *case-structured* tasks in addition to sequential ones. A case-structured task is identified by a bubble that controls activity in the network. The bubble has an input dataflow that can be one of several different types, each of which initiates different processing. The function of the bubble is to act like a switch in a train yard: it gets one piece of data, decides which type it is, and then activates one of a set of mutually exclusive processes,

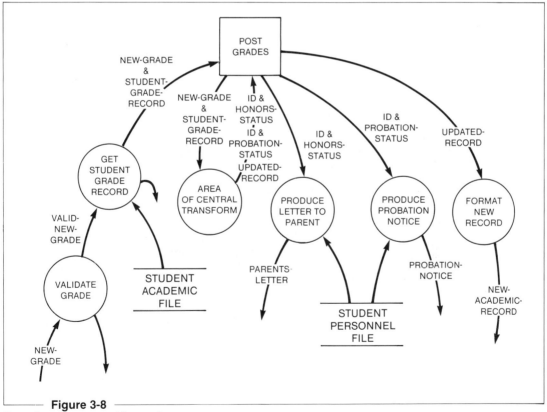

Figure 3-8

Dataflow diagram with new boss at top.

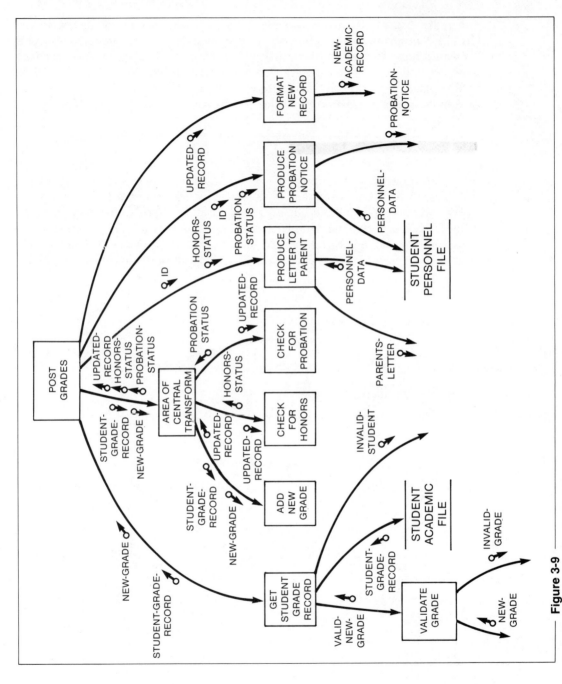

Figure 3-9

Introduce structure chart graphics.

depending on the value of the input dataflow. This controlling bubble is called a *transaction center;* the strategy we use to derive a first-cut structure chart from a dataflow diagram containing a transaction center is called *transaction analysis.*

Frequently menu-driven online systems have transaction centers that control the invoking of subsystems. The user keys in a command. The transaction

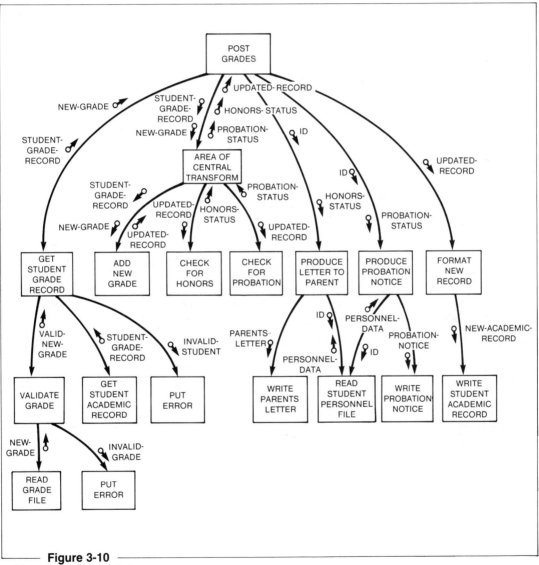

Figure 3-10

Add file interface modules.

center interprets the command and then invokes the routine the user requested. A word processor, for example, is a special online computer system. It has all the necessary components: hardware (a keyboard, a visual display unit, computer memory, and offline storage); programs (they come on diskette or tape, and the user loads them when needed); data (the user enters text via the key-

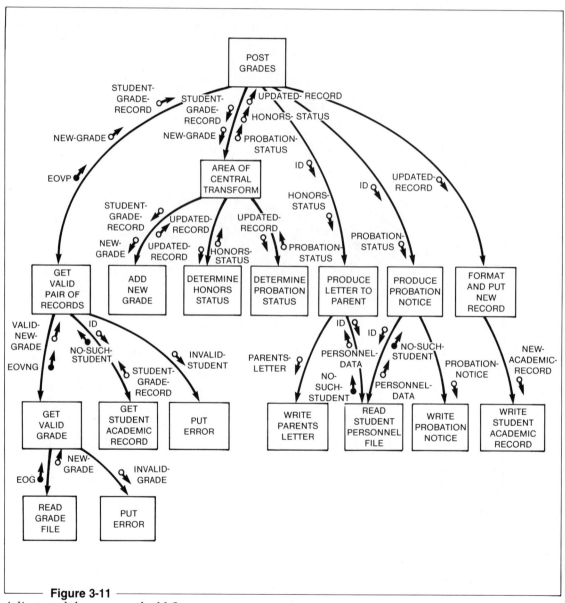

Figure 3-11

Adjust module names and add flags.

board); procedures (documented in user manuals, visual prompts on the screen, training manuals); and people (the hands-on user).

Look at the word-processing menu illustrated in Figure 3-13. The user can perform one of three activities from the menu: edit a document that is stored on diskette, create a new document, or print a document. The dataflow diagram for this simple word processor appears in Figure 3-14. (Notice the dashed lines. We will explain them shortly.)

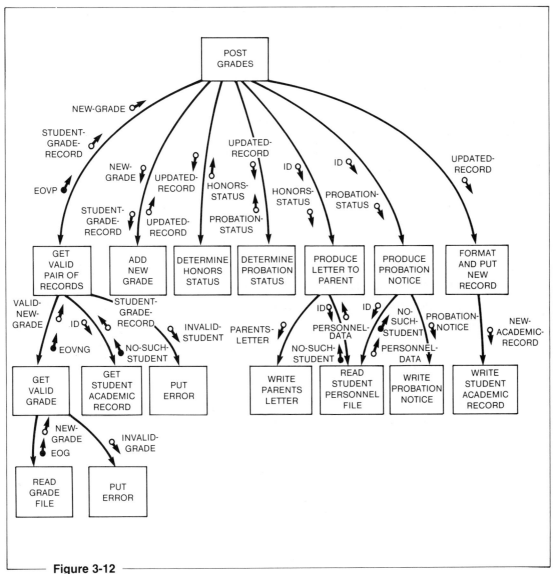

Figure 3-12

First-cut structure chart for POST GRADES.

This is an example of a *case*; there are three possible valid menu choices. One process out of a set of mutually exclusive processes will be invoked, depending on the value of MENU-SELECTION. The transaction center is the bubble called INVOKE SUBSYSTEM. It accepts MENU-SELECTION, decides which of the three options the user wants to perform, and then invokes that particular option (if the user has requested an invalid menu selection, INVOKE SUB-SYSTEM issues an error message). Each of the "working" bubbles—EDIT DOCUMENT, CREATE DOCUMENT, and PRINT DOCUMENT—is responsible for getting its own input data. They receive no data, only orders, from the controlling bubble, INVOKE SUBSYSTEM. The dashed lines indicate that control, not data, is passed from one process to another. Some of the dataflows needed by the process bubbles are the same, such as DOCUMENT-NAME, and some of the activities inside the bubbles might be identical. But each bubble is treated as an independent unit, processing only its own dataflows, unaware of any other bubble's activity.

Deriving a first-cut structure chart from a dataflow diagram that contains a transaction center is really very simple. The transaction center becomes the boss; turn the dataflow diagram around until the boss is at the top, and voila! Steps 1, 2, and 3 are completed. See Figure 3-15. Steps 4 through 6 are exactly the same as they were in transform analysis. The steps are repeated here to refresh your memory.

Step 1.   Select a dataflow diagram.

Step 2.   Identify the area of central transform.

Step 3.   Redraw the dataflow diagram with the boss at the top and all the other bubbles underneath.

Step 4.   Introduce structure-chart graphics: rectangles, data parameters, and module-call arrows.

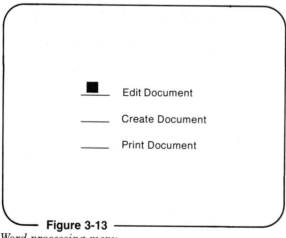

Figure 3-13
Word processing menu.

Step 5. Add read and write modules to access files.

Step 6. Change the module names in keeping with module-naming conventions; add flags.

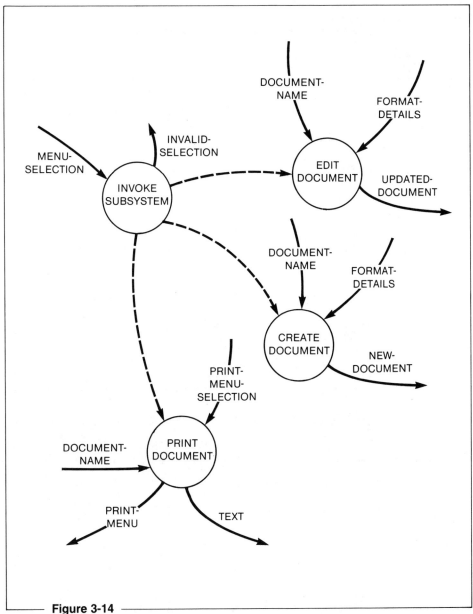

Figure 3-14
Dataflow diagram for Word Processor Control program.

Figure 3-16 shows the first-cut structure chart for the word-processing problem. Note the diamond symbol at the top. The diamond indicates that the boss will invoke no more than one of the subsystems at a time. Notice also that some modules (such as GET DOCUMENT NAME) are called by multiple bosses. Those modules clearly are utility modules. There may be many others, but we will not know for sure until we refine the structure chart (remember, it is only our first cut) and derive a far-more-detailed version. Once these utility modules are identified, they can be exploited. More about that later.

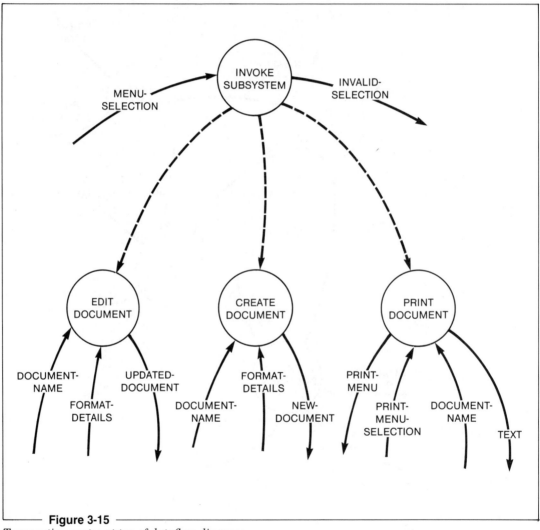

Figure 3-15

Transaction center at top of dataflow diagram.

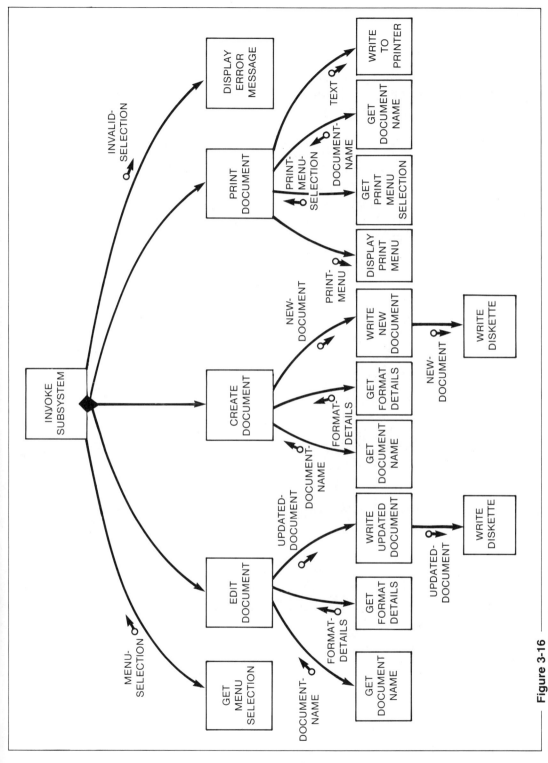

Figure 3-16
First-cut structure chart for word processor.

Transaction centers may be hidden inside a dataflow diagram bubble, too. Figure 3-17 shows part of a dataflow diagram for collecting town taxes. The bubble called Validate Tax Record is actually a transaction center: it handles four different types of tax records, although you cannot tell from the dataflow diagram.

The detailed activities inside VALIDATE TAX RECORD are illustrated in Figure 3-18. The first bubble, DISTRIBUTE TAX RECORD, checks to see what type of tax record is coming in, and then sends it on one of four conveyor belts to be validated (any that cannot be identified are rejected immediately). You see, each type of tax record is unique. So each type is validated a little differently from the others. Good tax records, no matter what type, are then passed to PRODUCE TAX STATEMENT.

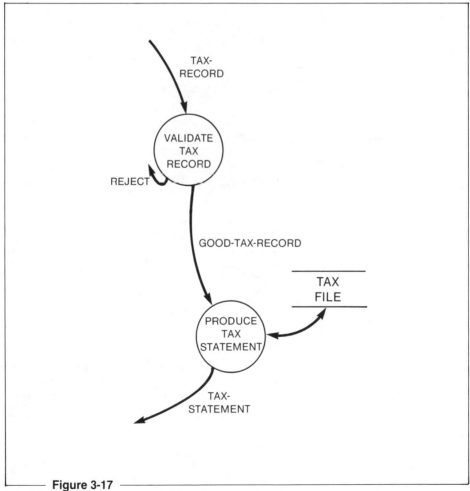

Figure 3-17

Partial dataflow diagram with hidden transaction center.

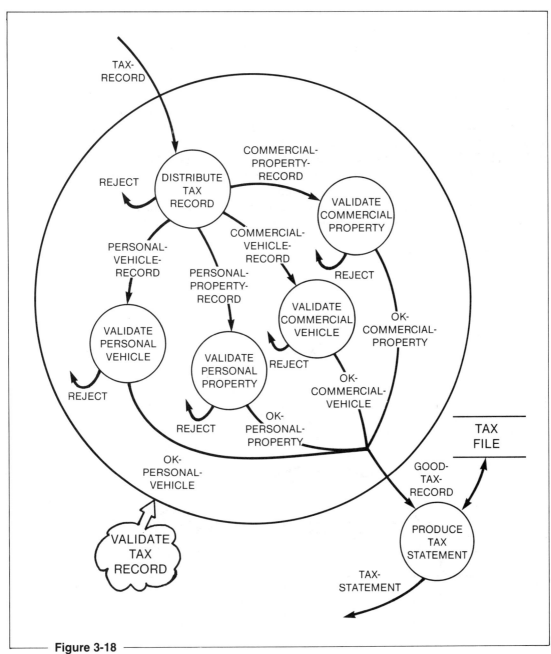

Figure 3-18
Transaction center.

Figure 3-19 shows part of a first-cut structure chart. Notice that VALIDATE TAX RECORD is innocent-looking enough. But Figure 3-20 illustrates the first level of modules beneath VALIDATE TAX RECORD: each module handles a different type of tax record, or transaction type. VALIDATE TAX RECORD, like its namesake on the dataflow diagram, simply distributes tax records to the appropriate validation modules.

It is at the next level (Figure 3-21) that we begin to see some utility modules. This level validates individual fields within a record. Some fields are common to several tax-record types—for example, DATE-ACQUIRED and VEHICLE-WEIGHT. The utility modules we have identified will eventually be coded only once and called each time they are needed, saving time and effort during implementation, and assuring some amount of consistency.

At the very bottom of Figure 3-21 are modules with stripes down the side, indicating that they are library routines already, or will be when the system

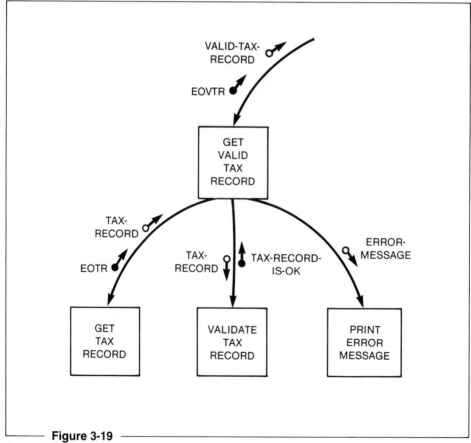

Figure 3-19

Part of first-cut structure chart.

is built. They are useful not only to this system, but also to other systems within the installation. They are VALIDATE STATE CODE and VALIDATE GREGORIAN DATE.

The important thing to remember is that, at the first level below the transaction center, the modules are separated according to transaction types, not common fields or other common processing. In the event that one type of tax record is someday eliminated altogether, updating the system will be easy: just delete the module dealing with that record type.

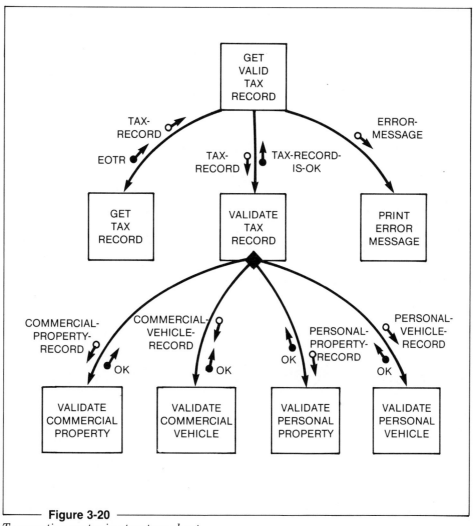

Figure 3-20

Transaction center in structure chart.

78

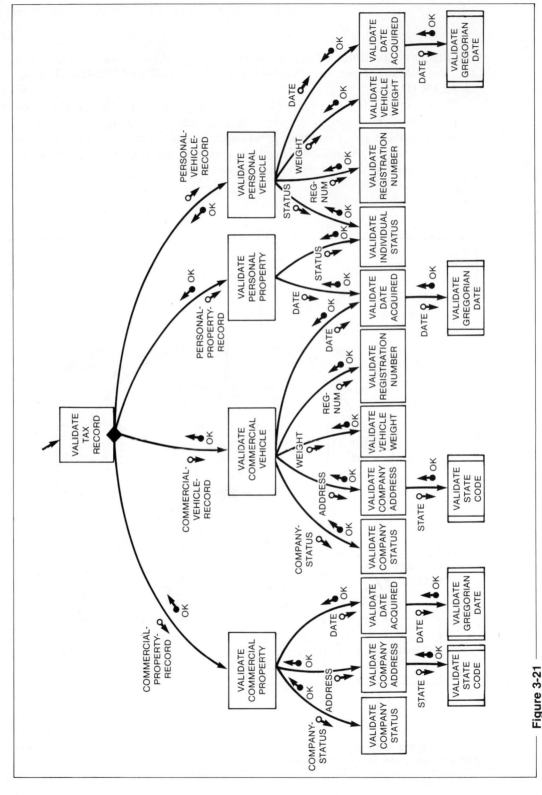

Figure 3-21

Factored modules.

CHOOSING YOUR APPROACH

Transform analysis and transaction analysis are frequently used together in producing a structure chart from a dataflow diagram. Pick the tool that best suits the problem. If you are dealing with an online system, you will probably start with transaction analysis to get the overall shape of the structure chart, then perform transform analysis on each transaction-handling process individually. That was the approach we took in the word-processing example.

Sometimes a transaction center is buried in the dataflow diagram, as it was in the tax-record validation example. In such a case, we derive the first-cut structure chart using transform analysis, then use transaction analysis on that part that deals with a transaction center.

Of course, if you are not sure, try both methods on the same dataflow diagram. It takes only a few minutes to rough out a first-cut structure chart. Draw two and throw away the one you don't like (paper is relatively cheap, and the exercise is hardly a waste of time!)

In fact, when it comes to deriving a first-cut structure chart, be prepared to make several false starts. The most skilled designers create, repair, massage, and discard several structure charts before they hit on one that shows promise. Good ideas seldom emerge fully developed from the mind of a human being. So be prepared to use a lot of paper.

SUMMARY

Dataflow diagrams are the basis for program designs. We derive structure charts from dataflow diagrams by using transform analysis for sequential tasks and transaction analysis for case-structured tasks.

The dataflow diagram we use must represent one single program. A program is a group of computer instructions that is loaded into the computer all at once and then executed. We isolate one program from another by examining online and batch application requirements, by eliminating functions that can be handled by commercial software, by introducing reports and intermediate files imposed by auditors for control over the system, by separating functions by the frequency with which they are executed, and by implementing functions on different machines.

Having isolated each program within the system, we can derive structure charts from the resultant dataflow diagrams.

Transform analysis and transaction analysis are strategies used to derive a first-cut structure chart from a dataflow diagram.

The steps we perform in transform analysis are—

1. select a dataflow diagram for one program;

2. identify the area of central transform, or the "boss";

3. copy the dataflow diagram with the boss bubble at the top and all the other bubbles underneath it;

4. change bubbles to rectangles, change dataflow arrows to module call arrows, and add data-parameter symbols;

5. add read and write modules for file access;

6. adjust module names in keeping with module-naming conventions; add flags.

A transaction center is a bubble in a dataflow diagram that handles a case. The bubble examines its input dataflow and then, based on that, decides which one of several mutually-exclusive processes to execute. Sometimes a transaction center controls the activity of a system, so it appears at the top of the structure chart. Sometimes it is hidden inside a bubble and is not discovered until transform analysis is complete.

We illustrate a transaction center on a structure chart with a diamond symbol. The diamond shows that the boss module will call only one of the worker modules below it at a time. Each worker module handles one and only one type of transaction. At a lower level, there are often utility modules—modules that handle functions common to the processing of several transaction types.

The first-cut structure chart is not a finished product. In Chapter 4, Refining the Structure Chart, we will learn how to identify flaws in the first-cut structure chart, and how to refine it into a finished product to form the basis of our programming specifications.

KEY WORDS

Area of central transform   The process bubble or bubbles in a dataflow diagram that deal with the most abstract input dataflows and produce the most abstract output dataflows.

Batch processing   Collecting and processing data in groups. See also *online processing*.

Case-structured task   A task in which one and only one process is selected from a set of mutually exclusive processes. The selection depends on the type of input dataflow. See *transaction center*.

First-cut structure chart   The output of either transform analysis or transaction analysis. A first-cut structure chart usually needs of further refinement before it is an adequate program model.

Online processing   Processing data as

soon as they are made available. See also *batch processing*.

Program   A set of computer instructions that is loaded by the operating system all at once and then executed.

Sequential tasks   Tasks that are executed unconditionally. See *case-structured tasks*.

Transaction analysis   A strategy for deriving a structure chart from a dataflow diagram that illustrates a *case-structured task*.

Transaction center   The name given to the process bubble that controls the selection of a process in a case-structured task.

Transform analysis   A strategy for deriving a structure chart from a dataflow diagram that illustrates sequential tasks.

EXERCISES

1. Using the dataflow diagram in Figure 3-1, do Steps 1 through 4 of transform analysis, selecting CHECK BOOK OUT as the boss.

2. What are the differences between the structure chart you derived in Exercise 1 and the one in Figure 3-3?

3. At what step in transform analysis did you realize that CHECK BOOK OUT is probably not the "best" boss?

4. Using the dataflow diagram in Figure 3-E4, RESPOND TO BOOK INQUIRY, derive a first-cut structure chart.

5. Using the dataflow diagram in Figure 3-E5, RESERVE BOOK, generate a first-cut structure chart.

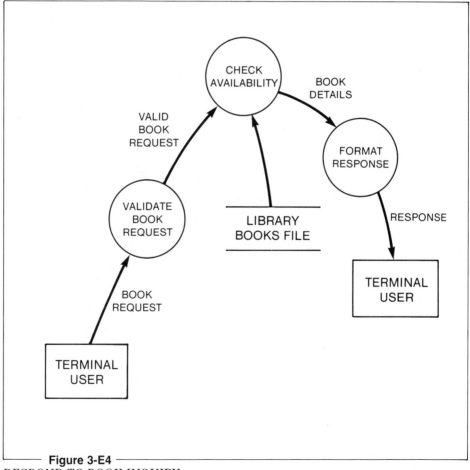

Figure 3-E4

RESPOND TO BOOK INQUIRY.

82

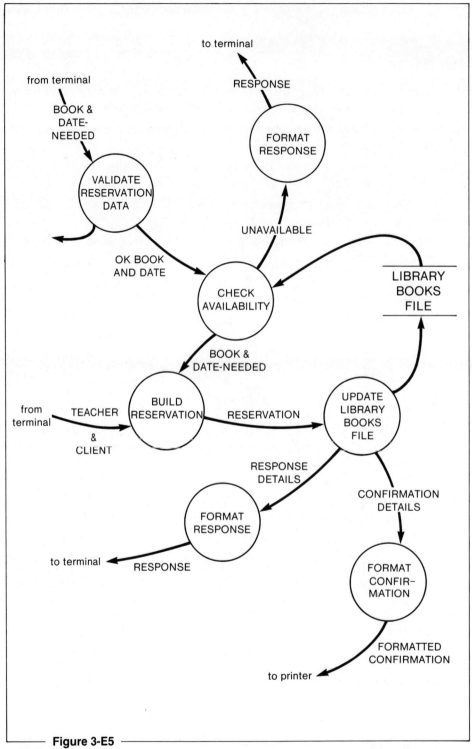

Figure 3-E5
RESERVE BOOK.

6. Using the dataflow diagram in Figure 3-E6, CONTROL LIBRARY SYS-TEM, derive a first-cut structure chart.

7. Combine your answers to exercises 4, 5, and 6 into one structure chart (you'll need a large piece of paper and a sharp pencil). Take note of any modules that appear multiple times on the chart.

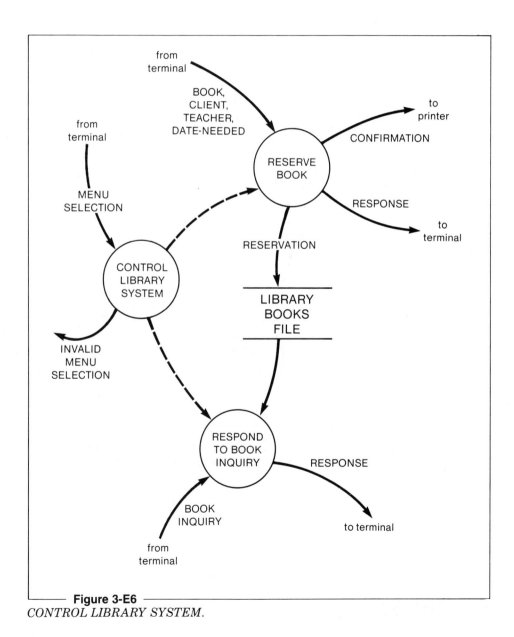

Figure 3-E6
CONTROL LIBRARY SYSTEM.

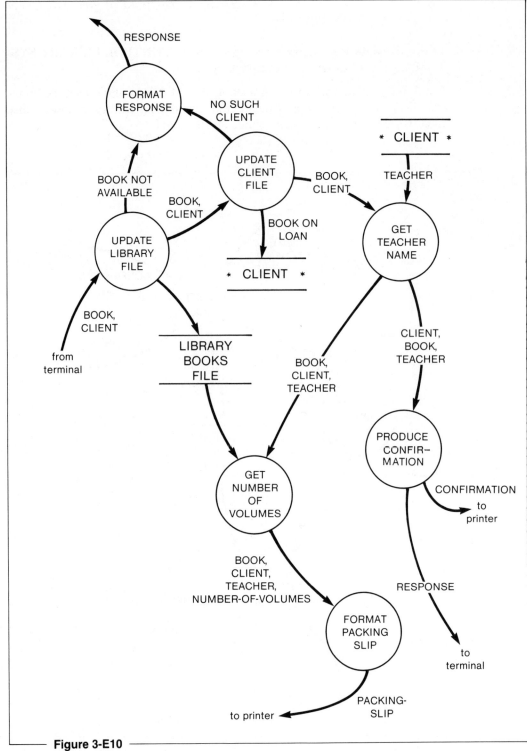

Figure 3-E10
CHECK BOOK OUT.

8. Play with the structure chart you produced in Exercise 7. Factor out any generally-useful functions you can think of. Rejoice when you find a utility module (for extra credit, of course).

9. Compare your structure chart with someone else's. Note similarities and differences. Combine the best aspects of both to create an even better structure chart.

10. Using the dataflow diagram in Figure 3-E10, CHECK BOOK OUT, derive a first-cut structure chart.

11. Add the structure chart from Exercise 10 to the structure chart you produced in Exercise 7. (It's getting pretty big, isn't it?)

CHAPTER 4

Refining the Structure Chart

When you finish this chapter, you will be able to improve a first-cut structure chart by examining these aspects of the structure chart:

- cohesion
- coupling
- span of control
- module usefulness
- module size
- common storage areas

INTRODUCTION

In the preceding chapter, we discovered how to derive a structure chart from a dataflow diagram. We noted that the resulting structure chart was a good start, but that it needed further refinement before it could become the basis for program specifications. In this chapter, we see how to critically assess a structure chart in order to find and correct flaws, so that the program based on the design will be easy to develop, understand, and modify.

First we examine the characteristics of *black boxes,* and then relate them to modules on a structure chart. Then we take a closer look at structure charts, searching for clues that point to design weaknesses. We may discover that a module has less than ideal *cohesion.* We learn what clues point up problems with *coupling* between modules. We adjust a module's *span of control* in order to limit the number of interfaces with which the programmer must ultimately deal. We examine the *usefulness* of each module. We even consider a module's *size*—that is, the number of lines of code it is likely to contain when it is coded. As we find flaws, or areas for improvement, we learn how to eliminate them and strengthen the design of the program.

We encounter trade-offs, of course. Changing a structure chart in order to eliminate one design flaw can sometimes mean introducing a different flaw. Fortunately, creating a perfect structure chart is *not* our goal; we simply want to design a program that will be relatively easy to develop, relatively easy to understand, and relatively easy to maintain.

We would like to design programs from modules that are cohesive, loosely coupled, have a limited control span, are generally useful, and are of moderate size. We may not be able to achieve all these qualities in all modules but, if we pay attention to these characteristics, we increase our chances of designing programs that are easy to develop, understand, and maintain.

BLACK BOXES

GENERAL CHARACTERISTICS

We call an assembly that can be used without knowing how it works a *black box.* A car, for example, is a black box for most people. Most of us know how to operate one, know how to perform some maintenance functions, and know that when the tank nears empty we have to fill it up. But many of us do not know how internal combustion engines work. One nice aspect about a car is that we don't have to know how it works in order to use it.

We know what a car does (it gets us from point A to point B); we know what we have to give it so it will run (gasoline, oil, and air in the tires), and we know the procedures for making it go (turn the ignition key, use the pedals and the shift lever, turn the steering wheel). Many of us have been doing this for years, but we probably will never crawl under the hood to figure out what is really going on inside, because we don't have to.

Our world abounds with black boxes: do you know exactly how your wristwatch works? Or your television set? How about your stereo or your camera? In fact, we constantly use black boxes—objects that perform a function when they are called upon to do so—without really knowing (or caring, probably) how they go about doing their thing. Here are some characteristics of black boxes.

1. One does not have to understand how a black box works in order to use it.

2. Black boxes are independent of each other, except at points where they interface. This makes it easy to add new features to the system without adversely affecting the performance of the rest of the system.

3. A black box can itself be composed of other black boxes. This helps to hide details, and thus to hide the complexity of the system from the user.

4. Problems with the system can usually be traced to one black box; the offender can then be repaired or replaced.

5. If one must become an expert on the operations inside one black box, it is relatively easy to do so.

Let's return to the automobile for an illustration.

Just because the average car owner has little interest in how an automobile engine works does not mean that no one does: consider the mechanic who does service and repair work on a car. Many service stations even have specialists on their staffs. One might take care of electrical systems, one might work on engines, and another might do body and chassis work: three individuals who specialize in three separate components. Just as the car itself is a black box, it is made up of interfacing black boxes.

The carburetor, for example, performs a function: it mixes gasoline with air to produce a vapor. If carburetor problems arise, the engine specialist is called upon to fix them. After studying the problem, he might elect to repair the defective carburetor itself, or he might choose to replace it with a new carburetor. In either case, the mechanic is aided by the fact that the carburetor is a black box. If he must replace it, it is relatively easy to remove the defective one and install a working one. If he fixes the original, then at least all the pieces of the carburetor are right there in one package. One could become a real expert in carburetors if one had to. And notice that the electrical specialist and body specialist are not involved at all with carburetor problems; nor does the engine mechanic affect either the electrical system or the paint job (we hope) by working under the hood.

That is because another characteristic of black boxes is that they are independent of each other, except at the points where they interface. The engine is fastened to the chassis, and the spark plugs produce an electrical spark that ignites the vapor produced by the carburetor. But they are still relatively independent of one another.

Installing a stereo system in an automobile is essentially introducing another black box into an already complex system. To the extent that the stereo installation interfaces with other systems in the car (the electrical system, for example, and the body of the car), there is some impact on the whole car system. Yet the impact is limited in scope, and it exists only at the points where an interface occurs. The performance of the car is not changed by installing a stereo. And if problems ever arise with that stereo, they are probably not caused by a dent in the body or a clogged air filter.

Because the dependencies of one black box on another are known and defined, black boxes can be easily installed, replaced, and repaired without adversely affecting the rest of the system. And if the number of interfaces is kept to a minimum (two wires rather than twenty), then coupling two systems together is clean and easy to understand.

MODULES: BLACK BOXES FILLED WITH CODE

In a program, the black boxes are called modules. We saw in Chapter 2 that a structure chart is a picture of interfacing modules. The characteristics that make black boxes attractive apply to modules as well.

1. Modules can be used without having to be understood internally by the user (but their functions must be known).

2. Modules are independent of each other, except where they interface.

3. They can be further subdivided into smaller modules, thereby reducing complexity by hiding details.

4. Program problems can be easily traced to a faulty module that can be diagnosed, replaced, or repaired.

5. They are easy to understand (relatively speaking) if one must understand them.

Our first-cut structure chart is our design starting point. Although every module may not yet be a perfect black box, our goal is to make as many of them black boxes as possible.

The topics we will now address are *cohesion, coupling, span of control, module usefulness, module size,* and *common storage areas*. They will guide us closer to our goal of good program design.

COHESION

We said earlier that a car is a black box, made up of several interfacing, but smaller, black boxes—such as the carburetor, the fuel pump, and the stereo.

Each of these items is a complete functioning unit, almost independent of the rest of the car (complete independence is, of course, impossible: the carburetor depends on the fuel pump for some of its input, and the stereo will not work unless it interfaces with the electrical system).

One characteristic common to the black boxes is that each one contains all of the mechanical, electrical, and electronic parts to perform its unique function. Except for the known and defined interface to another black box, each is a complete unit; and each black box serves only one function.

When a black box contains all the things it needs to accomplish a single function, we say that it has *cohesion*.

Modules, because they also are black boxes, share this characteristic. If all the instructions inside a module accomplish a single function, then the module has high cohesion. To discover why this is so important, we return to the automobile.

By gathering all the components of the stereo into one package and physically isolating it (a very cohesive package), we make it *easy to install* (no other parts of the system are affected except those that interface with the stereo), *easy to test and fix* (if it doesn't work we know the problem lies either within the stereo itself or in the interface with the electrical system), and *easy to replace or remove* (the whole stereo can be disconnected and replaced with a new one or removed altogether).

The same is true of programs we design. If each module in the program contains all the instructions needed for a single function, then it is easy to build, test, and debug; easy to install in the program to see how it functions with interfacing modules; easy to replace if its function changes; and easy to remove if the need for its function is eliminated. The reason is simple: we can deal with each module without having to pay any attention to any other modules in the program, except for the ones with which it interfaces.

When we study cohesion, we isolate each module on a structure chart and examine the module's name, its boss and workers, and the flags it receives from its boss. When we discover a module that does not complete a single well-defined function (it might contain pieces of several functions, only part of a function, or several complete functions), we improve it and thereby improve the design of our program.

First, let us look at some modules that exhibit good cohesion. Each of the modules in Figure 4-1 performs a single well-defined function; each one has a strong specific name; and each one receives only the data it needs to do its job, and returns only the answers its boss wants. These are highly cohesive modules.

By contrast, we are going to spend the remainder of this section on cohesion looking at less-than-ideal modules. We will learn how to identify weaknesses, and how to fix them and improve our design.

The structure chart gives us three areas to study for clues to potential cohesion problems:

1. the module name;

2. the organization of the modules;

3. the flags passed between modules.

MODULE NAMES

COHESION CLUE 1:   Module name contains connective (and, or, then, commas).

PROBLEM: Module performing multiple functions.

SOLUTION: Decompose module into several single-function modules.

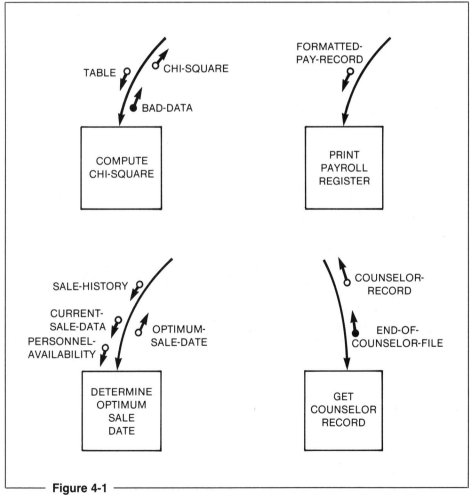

Figure 4-1

Highly cohesive modules.

Here are the names of some multiple-function modules:

Compute Tax Payment or Credit-Union Deduction

Print Counselor Caseload and Quarterly Report

Display Mean or Median or Mode

Get Next Record Then Validate Account-Number Field

Calculate Quarterly Dividend, Print Dividend Check, Get Next Client Record

Figure 4-2 shows how each of these modules would look on a structure chart. In every case, the module contains instructions to accomplish more than a single well-defined function. And in every case, we can improve the cohesion of the module by breaking it into multiple single-function modules.

Compute Tax Payment or Credit-Union Deduction becomes
　　Compute Tax Payment
　　Compute Credit-Union Deduction

Print Counselor Caseload and Quarterly Report becomes
　　Print Counselor Caseload
　　Print Quarterly Report

Display Mean or Median or Mode becomes
　　Display Mean
　　Display Median
　　Display Mode

Get Next Record Then Validate Account-Number Field becomes
　　Get Next Record
　　Validate Account-Number Field

Calculate Quarterly Dividend, Print Dividend Check, Get Next Client Record becomes
　　Calculate Quarterly Dividend
　　Print Dividend Check
　　Get Next Client Record

Why is this an improvement? Because all the tasks pertaining to one activity are isolated into one physical place; then developing, understanding, and maintaining that activity are relatively easy. Also we have created several modules that are potentially useful to other programs or systems. Another programmer probably will have no use for a module called Calculate Quarterly Dividend, Print Dividend Check, Get Next Client Record. However, Calculate Quarterly Dividend may be very useful without all that excess baggage, and Get Next Client Record is sure to be popular in many applications.

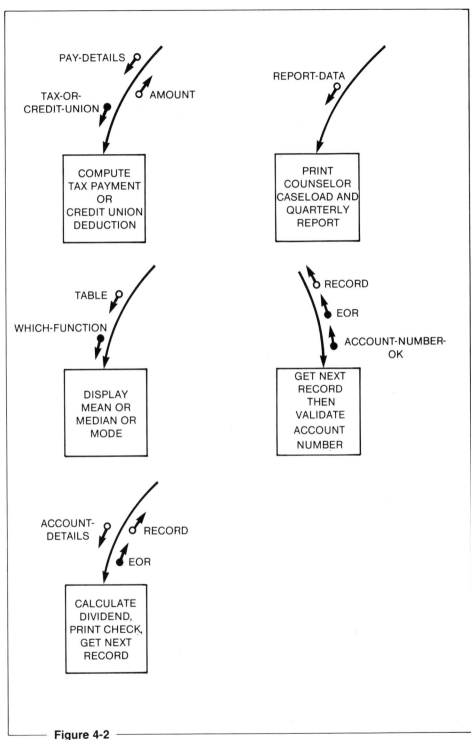

Figure 4-2
Multiple function modules (needs improvement).

COHESION CLUE 2: Nonspecific module name.

PROBLEM: Module performing multiple functions.

SOLUTION: Decompose module.

Sometimes a module name is deceiving: it has the required verb and object, but one of them (sometimes both of them) is weak, or nonspecific. If you find a module with a weak verb (DO, PROCESS, TERMINATE) or a weak direct object (NAMES, PARAMETERS, DATA), write down the instructions you would expect to find in the module, to get a feel for what the module is doing. Frequently, the reason a designer is unable to think of an accurate module name is that the module is doing many unrelated things.

Figure 4-3 shows a module called INITIALIZE WORK SPACE. The pseudocode* for INITIALIZE WORK SPACE was written as follows:

```
Open files
Print page heading
Move zero to sub, record count
Move 1 to page number
Get first record
If record type = '1', move 'X' to flag
```

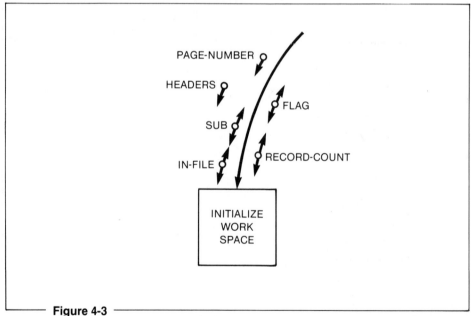

Figure 4-3

Nonspecific module name (needs improvement).

*Pseudocode is intended to show what the module does, without using a formal programming language.

The designer realized that the instructions inside INITIALIZE WORK SPACE were not at all related, so she wisely decomposed it, placing the moves, opens, and prints where they belong: in other modules.

For example, one module on the same structure chart was called READ TAX FILE. It made more sense to have that module open its own file the first time it was called. By moving the open statement for the tax file into the same module that read the file, the designer kept related things together. Consequently, she improved the design. Of course, she also had to include instructions in that module that would ensure that the file was opened only once.

Look carefully at each module's name. Weak verbs can be clues to potential problems.

COHESION CLUE 3:   Assembly-line name.

PROBLEM: Module has a poor name.

SOLUTION: Rename the module.

Before you take a red pencil and a pair of scissors to a structure chart when you see a less-than-perfect module name, consider this: maybe the author (you) had a mental block while naming the module and misnamed it. Although the name looks like an assembly line—that is, it lists a series of steps that are done one after the other on some piece of data—maybe only one function is actually being completed. Here is a rather long module name, chock full of commas:

Get Time Sheet, Calculate Gross Pay, Calculate Net Pay, Print Check, Print Pay Stub, Update File

Now there's a module that looks like a candidate for the chopping block, but wait! All of those steps contribute to the completion of one task: Produce Payroll. Of course, if one module contained *all* of the code to do all of the steps, it would undoubtedly be too large (see the discussion of module size later in this chapter). An appropriate structure chart for Produce Payroll shows that the boss module simply controls the sequence in which the worker modules are called (see Figure 4-4).

Every module name must be carefully studied. If there is any doubt in your mind that the module is completing a single well-defined function, then jot down the instructions that are likely to appear in the module. If you find multiple functions, factor them out. Each new module then will have a good name— one made up of a strong active verb and a singular specific direct object.

MODULE ORGANIZATION

In order to evaluate module cohesion, we also study the physical organization of the modules on a structure chart. We examine who reports to whom. Sometimes we find workers that are simply reporting to the wrong boss. Our clue is that the function the worker performs is not one its boss needs in order to perform its function.

COHESION CLUE 4:   Worker's function does not help its boss.

PROBLEM: Worker reporting to the wrong boss.

SOLUTION: Physically move the worker on the structure chart.

Consider a program that prints two reports: a list of employees with some summary information, and a list of the names of each employee's dependents. The dependents field is part of the employee record. With this in mind, we could design the program so that the dependents field is printed as soon as the employee record is read. Other employee information needed for the summary report is passed up for further processing. Figure 4-5 illustrates this approach.

Let's interview each of the modules reporting to GET GOOD EMPLOYEE RECORD.

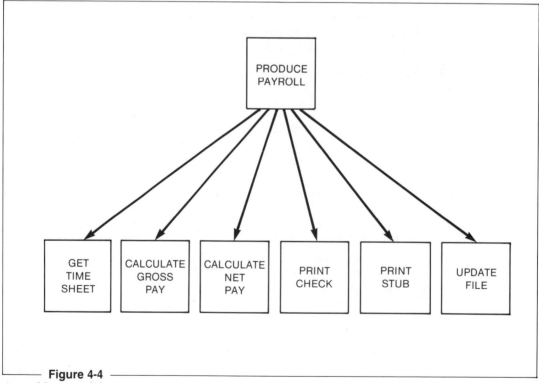

Figure 4-4

A good boss module.

US: Tell me, GET EMPLOYEE RECORD, does your function contribute to the overall function of your boss, GET GOOD EMPLOYEE RECORD?

MODULE: Yes, indeed, because in order to get a good record, my boss needs me to get the next record.

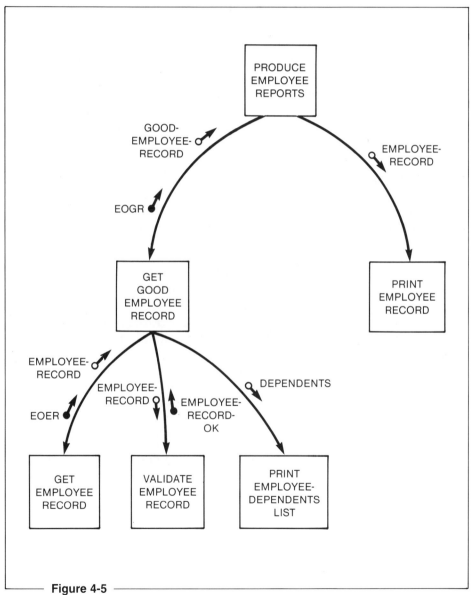

Figure 4-5

Module reporting to wrong boss (needs improvement).

US: And how about you, VALIDATE EMPLOYEE RECORD, is your function necessary for your boss to do its job?

MODULE: Of course. My boss gives me a copy of any old employee record, and I check it out to see if it's any good. I send back a flag to let my boss know my findings.

US: It's your turn, PRINT EMPLOYEE-DEPENDENTS LIST. Tell us how your function contributes to your boss's job of getting a good employee record.

MODULE: I wish I could do that. I have an important job to do, but I don't think I belong here. I mean, GET GOOD EMPLOYEE RECORD has to get a record and then validate it. I am not helping with that task one single bit. It just happens that the data I need is available down here. But between you and me, I think I belong somewhere else on the structure chart.

Which module in this example has weak cohesion? If you said GET GOOD EMPLOYEE RECORD, you are absolutely right! That module is really attempting to do two unrelated tasks: get a good employee record *and* produce a report. Yet its name specified only one of those functions. The solution is not to rename the module GET GOOD EMPLOYEE RECORD AND PRODUCE DEPENDENTS LIST, but to move the report-producing module to a more appropriate location in the structure chart.

To whom should PRINT EMPLOYEE DEPENDENTS LIST report? It appears that a more appropriate boss is PRODUCE EMPLOYEE REPORTS. After all, printing one of the reports is clearly making a contribution to the overall task of printing both of the reports. Figure 4-6 shows a revised structure chart for this example. (In Figure 4-6 and in some other illustrations as well, labels have been intentionally omitted from parameters for simplification. All parameters on a complete structure chart are labeled, of course.)

So the structure chart can reveal design flaws in module organization. It is very easy to discover errors by asking each module what it does to help its boss. If the answer is "nothing," then the module is a candidate for relocation.

FLAGS

The last set of clues to poor cohesion, and thus to design weakness, is found not in the modules themselves but in the flags that are passed between modules.

Flags, also known as switches or return codes, are both useful and necessary in many programming applications. Like so many other techniques, they can be used properly and in a controlled manner—or they can be misused, leading to unintelligible code. It is because of the misuse of flags that some programming standards manuals forbid the use of flags altogether! Why is it that the use of flags can evoke such an extreme reaction?

The reason lies in the fact that, as soon as flags are passed between modules, the modules are no longer black boxes. Both modules must know the possible flag settings and their meanings. Sometimes the only way to be sure is to look at a module's source statements. The result is that one module cannot be developed in complete independence from another one. Although this is not always a problem, there is a risk of passing flags that are unnecessary, whose names are misleading, or that serve multiple functions.

The solution to the problem is not to banish the use of flags, but to use them wisely and to eliminate any that are unnecessary. If we do use them, then we should name them accurately and should use one flag for only one function.

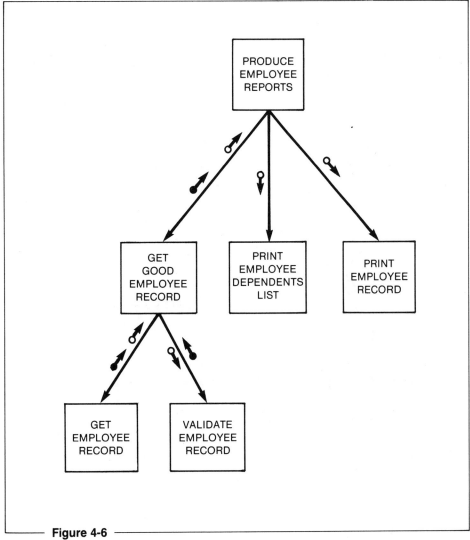

Figure 4-6

Module reporting to correct boss.

In order to determine if it is appropriate to pass a flag from one module to another, we must establish the function of the flag. Flags fall into three functional categories:

Reporting flag.   A reporting flag (sometimes called a return code) is passed from a worker to its boss. Its function is to report on a situation that has occurred. The boss might test the flag and take some action based on the results of the test. Names of reporting flags usually contain adjectives (see Figure 4-7).

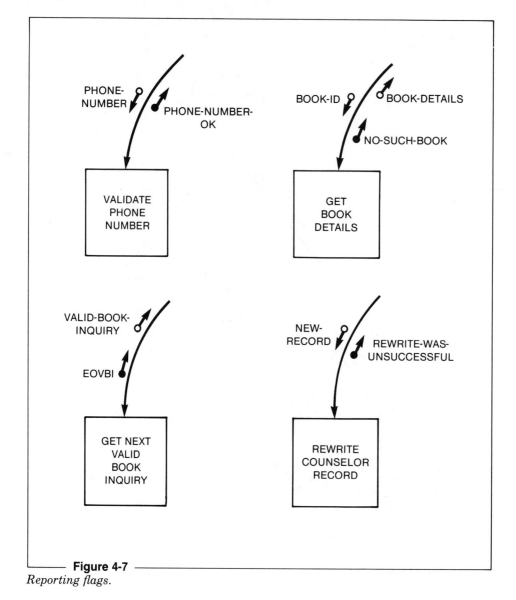

Figure 4-7
Reporting flags.

Activity flag. An activity flag is passed from a boss to one of its workers. Its function is to tell the worker module which of its several functions to execute [see Figure 4-8(a)].

Directive flag. Like a reporting flag, a directive flag is passed from a worker to its boss. The flag tells the boss what to do next. The name of a directive flag usually contains an imperative verb [see Figure 4-8(b)].

In this section on identifying modules with cohesion problems, we focus on activity flags. We examine directive flags and reporting flags later, in our discussion of coupling.

COHESION CLUE 6: Module receives an activity flag from its boss.

PROBLEM: Module is performing multiple functions, *and* it is not a black box.

SOLUTION: Factor single functions out of the module.

The name of a multiple-function module usually gives it away, but the activity flag is a sure sign of cohesion problems. The worker module is able to do

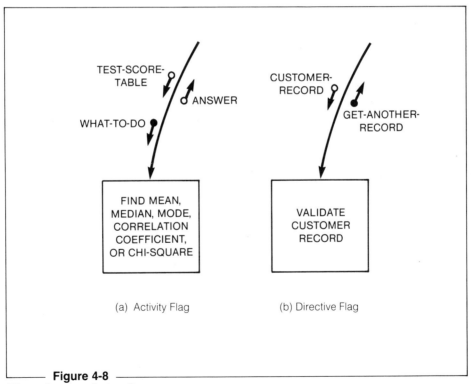

(a) Activity Flag   (b) Directive Flag

Figure 4-8

Flags showing design flaws.

several different (maybe even unrelated) functions; the boss decides which one it wants performed, and sends an activity flag to the worker. The flag is called an activity flag because the boss is telling the worker which activity to execute. Of course, the problem is that the worker does too many things.

Figure 4-9 shows part of a structure chart for a master-file-update program. The boss, UPDATE FILE, calls GET A RECORD to get either the next trans-action-file record or the next master-file record, depending on which one the boss needs. It sets the activity flag, WHICH-FILE, to 'M' if it needs the next master record and to 'T' if it needs the next transaction record.

The module name GET A RECORD looks innocent enough. It has a strong imperative verb and a singular specific direct object. Our clue to poor cohesion is the flag, WHICH-FILE. Here is the pseudocode for GET A RECORD:

```
if which-file-flag = 'M'
     read master file into next-record
          at end move 'Y' to eor-flag
else
if which-file-flag = 'T'
     read trans file into next-record
          at end move 'Y' to eor-flag
```

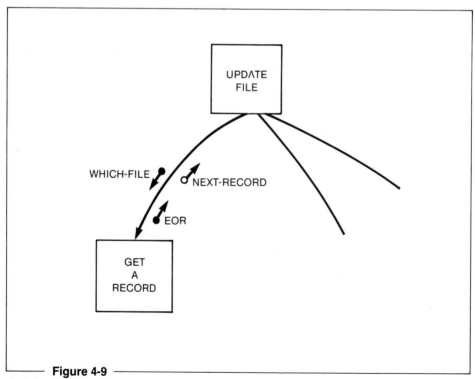

Figure 4-9

Activity flag (needs improvement).

```
else
      move spaces to next-record
endif
goback
```

The innocent-looking module is really performing two different functions (reading a record from the master file, and reading a record from the transaction file), and on that basis alone it can be improved by factoring the two functions into separate modules.

Another (more serious) problem is that a programmer who wants to use this module has to look at either the source code or program documentation for GET A RECORD to discover that 'M' means master and 'T' means transaction. (It could just as easily be 1 for master and 2 for transaction, or '*' for master and '%' for transaction.)

In order to improve the module, we will factor out the two functions into separate modules: one to read the master file, and the other to read the transaction file. The result is two highly cohesive modules, GET MASTER RECORD and GET TRANSACTION RECORD (see Figure 4-10).

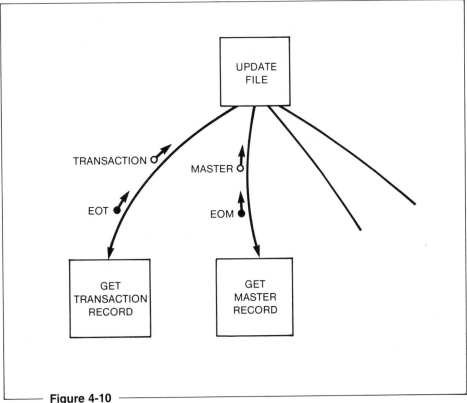

Figure 4-10

Activity flag is gone: Design is improved.

Now UPDATE FILE simply calls either GET TRANSACTION RECORD or GET MASTER RECORD. The activity flag, WHICH-FILE, has been eliminated. Furthermore, the two new modules, GET TRANSACTION RECORD and GET MASTER RECORD, are potentially more useful than the multifunction module they replaced. Maintenance, replacement, or removal of either of the two new modules can be done without affecting the other one. Program design has been improved as a result of this simple change.

COHESION: A SUMMARY

The term *cohesion* refers to how closely the instructions inside a module are related. A highly cohesive module is one in which all of the instructions contribute to the complete execution of a single well-defined function.

There are three clues to cohesion problems on a structure chart: module names, module organization, and activity flags.

MODULE NAME A module name that contains connectives indicates that the module is doing more than one complete function. A module name with a weak verb or a nonspecific object usually indicates that the module is performing more than one task. Improve the cohesion of the module by decomposing it into smaller functional modules.

MODULE ORGANIZATION A module that does not contribute to the overall function of its boss is probably reporting to the wrong boss. Improve the boss's cohesion by physically moving the worker to a more appropriate boss.

ACTIVITY FLAGS An activity flag is a flag that a boss passes to a worker, specifying which of the module's many functions is to be executed. A module that receives an activity flag from its boss is able to perform more than one function. Decompose the worker module. This eliminates the need for an activity flag, and it usually produces modules that are more generally useful.

COUPLING

The next design-refinement criterion we examine is called *coupling*. Coupling refers to the amount of dependence one module has on another. Modules that pass data are, of course, dependent on one another. Just as the carburetor needs the fuel pump to do its job, a boss module cannot perform its job very well without the worker modules that take some data from the boss and send answers back up to it. Bosses and workers depend on each other for data.

When we study coupling, we examine the number and type of parameters passed between two modules. (By contrast, cohesion refers to what goes on inside a single module.) When we discussed black boxes earlier in this chapter, we saw that modules connected by a few interfaces are easier to test, repair, and replace than modules sharing many complex interfaces with each other. We say that the good modules are *loosely coupled.* In our quest for good program design we will aim for systems built from highly cohesive, loosely coupled modules.

Our goal is to pass from one module to another only the amount of data that is absolutely necessary for the receiving module to do its job. This will result in modules that are loosely coupled.

As in the case of cohesion, we find that the structure chart contains most of our clues. We are going to focus on the parameters passed between modules, examining specifically

1. the number of parameters,

2. parameter names,

3. flags.

More often than not, a coupling problem is evidence of a cohesion problem. Coupling and cohesion problems go hand in hand: if you fix a coupling error, you almost invariably correct a cohesion problem at the same time. In fact, we already learned how some parameters—namely, activity flags—give us clues to cohesion problems. When we fix the cohesion problem, we usually eliminate the activity flag, thereby making the communicating modules less dependent on each other—that is, more loosely coupled. Let's look at some coupling problems and the clues we get from the structure chart.

NUMBER OF PARAMETERS

The first topic we examine is the number of parameters being passed between modules. Like Goldilocks, we do not want to pass too many or too few; we want to pass just the right number.

COUPLING CLUE 1: Module does not receive enough data to do its job.

PROBLEM: One or more parameters is missing.

SOLUTION: Make the missing parameter(s) available to the module.

Look at the structure chart in Figure 4-11. The module's function is to read a direct file and pass back a specific customer record, based on a key field. Can the module do its job? What is missing?

If you said the customer key field, you are correct. It is easy to fix the error. See Figure 4-12.

If there is any doubt at all that a module is receiving enough data to do its job, jot down the instructions you would expect to find inside the module. By

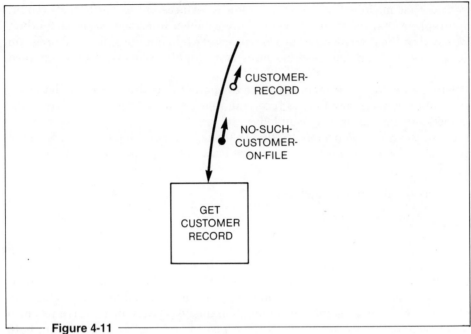

Figure 4-11

Module missing a parameter (needs improvement).

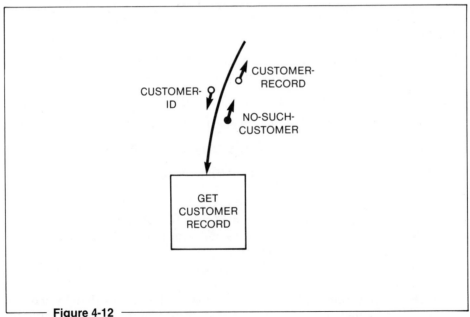

Figure 4-12

Parameter identified and included.

comparing what the module needs to what the module gets, you can identify any missing parameters.

COUPLING CLUE 2:   Module receives data it does not need.

PROBLEM A: A data structure containing both necessary and unnecessary data is being passed.

SOLUTION A: Extract the necessary fields and pass only them.

PROBLEM B: The module is acting as a go-between, accepting data it has no use for so that it can pass it to another module that does need it.

SOLUTION B: Make the module that needs the data responsible for getting it.

Figure 4-13 illustrates Problem A. The module UPDATE EYE-DISEASE MATRIX needs information about a client's eye condition to complete a matrix to be used for statistical evaluation. The two fields in the client record that the module needs are EYE DISEASE (in positions 48–87) and AGE AT ONSET (in positions 106–107). Of course, the client record contains dozens of other fields that are not needed by this module.

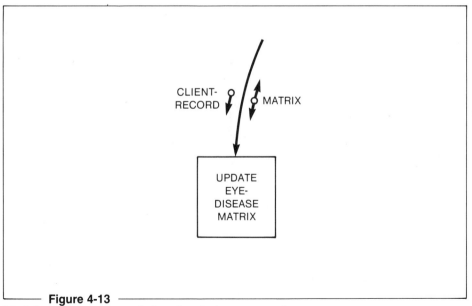

Figure 4-13
Module receiving too much data (needs improvement).

There are at least two problems here. (1) The module is subject to change if any change occurs in the structure of CLIENT RECORD, even though it does not use 90 percent of the data in the record. (2) Other modules that call this one must pass it a parameter that *looks* like a client record, being careful to place the eye disease in positions 48–87 and the age at onset in positions 106–107. In other words, any module that calls UPDATE EYE-DISEASE MATRIX must pass a dummy client record. What a foolish restriction!

From a non–data-processing standpoint, it may also be that this module has no legal right to the other parts of a client's record: name, place of employment, current salary, high-school grades, psychological evaluation, and so forth. Private or sensitive data must be protected, and access to it must be restricted. The solution is abundantly clear: pass to this module only the data it needs to do its job—namely, EYE DISEASE and AGE AT ONSET. See Figure 4-14 for an improved version of the module.

Now, no matter what CLIENT RECORD may look like, this module deals with only the two fields it needs to do its job. It is up to the calling module to extract those two fields and pass them down.

As noted above, the same clue might reveal another problem (Problem B), one in which a module receives data it has no use for, simply because some other module in the structure chart needs it. Figure 4-15 shows a structure chart containing a well-traveled parameter called ELIGIBILITY.

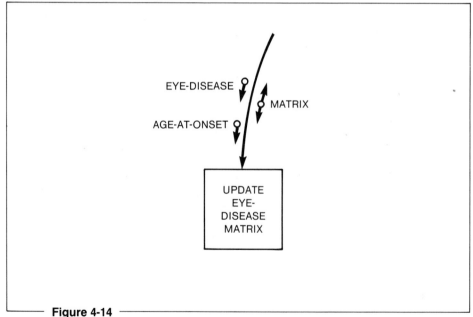

Figure 4-14

Unnecessary data eliminated: Design is improved.

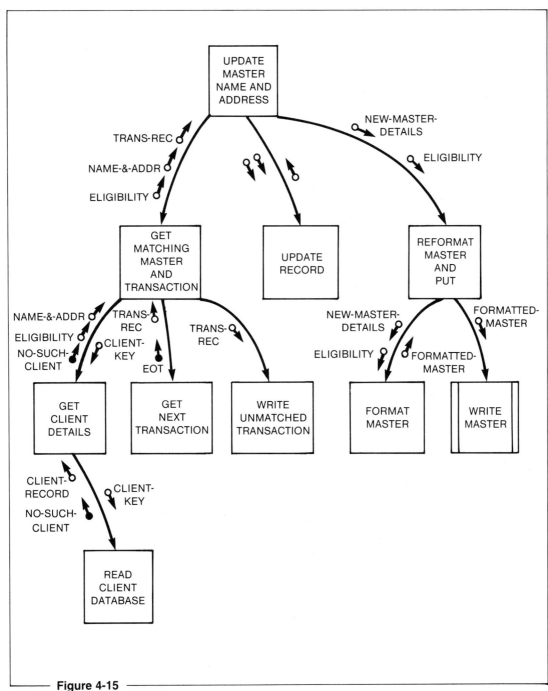

Figure 4-15

Parameter traveling too far (needs improvement).

ELIGIBILITY happens to be found in the client database, so the module GET CLIENT DETAILS grabs it along with other data fields and passes them all up to its boss. GET MATCHING MASTER & TRANS doesn't want it, so it passes it up to its boss, UPDATE MASTER NAME AND ADDRESS. That module has no need for it, so it passes it down the other side of the structure chart to REFORMAT MASTER & PUT, which passes it down to FORMAT MASTER, the module that really needs the field called ELIGIBILITY! It so happens that ELIGIBILITY indicates each client's source of funding, and FORMAT MASTER builds slightly different master records based on a client's source of funding. Well, if FORMAT MASTER needs it, let FORMAT MASTER get it from the client database, as illustrated in Figure 4-16.

Why is this an improvement? There are two reasons. First, ELIGIBILITY is no longer floating through the system where it is susceptible to inadvertent (or intentional) alteration by other modules. Second, the other modules in the system do not receive data for which they have no use, thereby decreasing need for maintenance changes if the format of ELIGIBILITY changes.

COUPLING CLUE 3:	Module deals with too many parameters.
PROBLEM:	Poor cohesion; module is performing multiple functions.
SOLUTION:	Decompose the module.

This coupling clue leads us to a cohesion problem as many of them do. You may be asking "How many parameters is too many?" There is no standard answer to that question. In some cases, two is too many, while in other cases 26 is not enough! A good designer examines the parameters passed between all modules and uses his head. Look at the structure chart in Figure 4-17.

It does not appear at first glance that this collection of parameters is needed to accomplish any one specific function. Furthermore, on closer examination, the module's name seems weak, a clue to possible cohesion problems. After writing a few lines of pseudocode, the designer decided that this module is really doing four different things, and changed the structure chart. Figure 4-18 shows the improved version.

Each of the new, smaller modules is highly cohesive and loosely coupled to its caller. Each one receives only the data it needs to perform a single well-defined function. As an added bonus, many of the new modules can be placed in a module library to be used by other programs.

Before you decide that a module has cohesion problems based on the number of parameters it receives, consider this: sometimes a designer starts off showing individual data elements, and then later collects them into a group item. A group item can be a natural data structure. You might be suspicious of the module in Figure 4-19.

Do not let the five input parameters fool you. Later on, the designer collected those fields together into a data structure called INTERNATIONAL PHONE

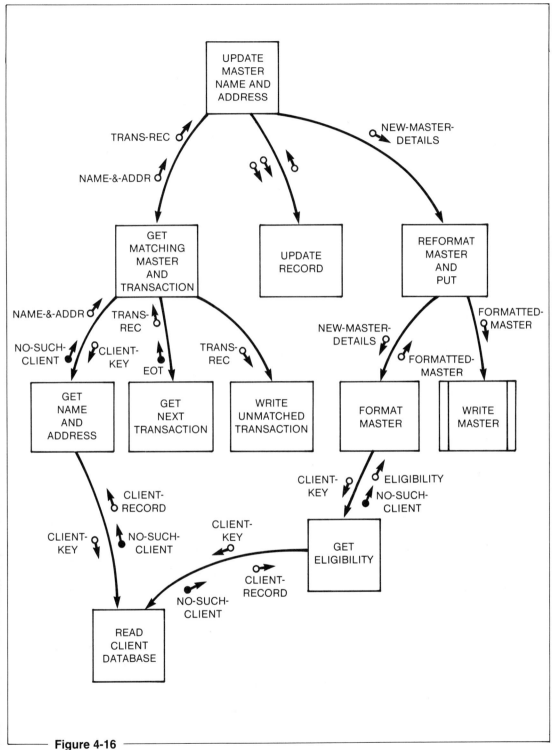

Figure 4-16

Parameter traveling only through modules that need it: Design is improved.

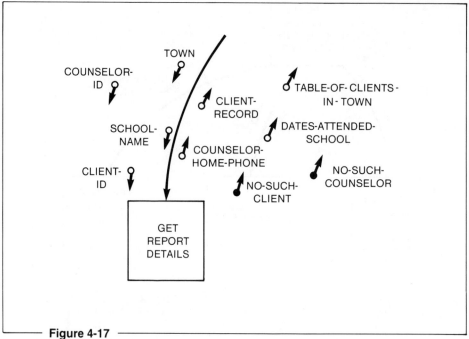

Figure 4-17

Too many parameters (needs improvement).

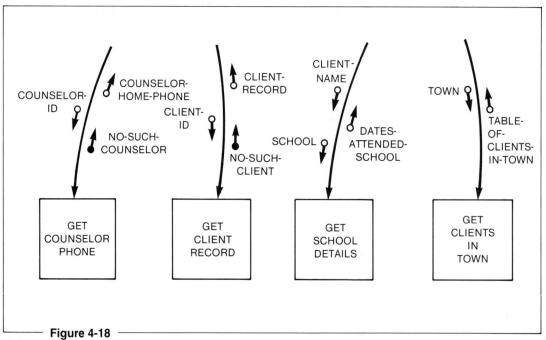

Figure 4-18

Cohesive modules factored out: Design is improved.

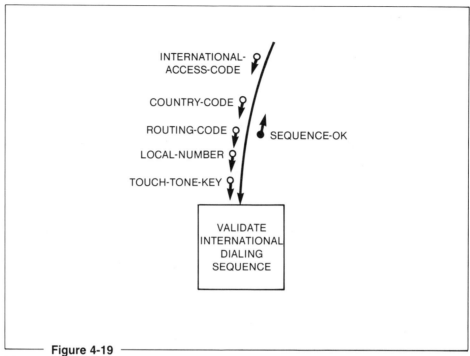

Figure 4-19

Too many parameters?

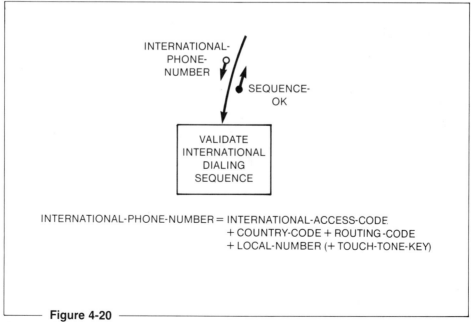

Figure 4-20

A natural data structure.

NUMBER. The final structure chart showed only one input and one output parameter (see Figure 4-20).

Arrays can be natural data structures. It is better to pass an array than it is to pass each element of the array as an individual parameter. When you see a large number of parameters, do not automatically assume that there are design problems. Try first to reduce the number of parameters by collecting *related* ones into natural data structures, either group items or arrays.

Warning: Do not fool yourself into thinking you have improved the coupling between modules by collecting *unrelated* data items into a group item called NAMES or DATA or PARAMETERS. Like an ostrich with its head in the sand, you are simply hiding from the problem, not fixing it.

PARAMETER NAMES

COUPLING CLUE 4: Data parameter has nonspecific name.

PROBLEM: It is an unnatural data structure.

SOLUTION: Decompose it into individual parameters.

Be very suspicious of parameters called AMOUNTS, NAMES, MINMAX, or other nonspecific words. Look at Figure 4-21. Any module that receives FIELDS and returns STUFF has a problem. First, none of the fields in Figure 4-21 could be found in the data dictionary. Second, the program specifications we will eventually write are supposed to be clear and unambiguous. Try to find a programmer who can distinguish between PARAMETERS and DATA!

If you cannot identify the parameters by name, then you do not yet have a clear understanding of the module's function. Establish that before you go on.

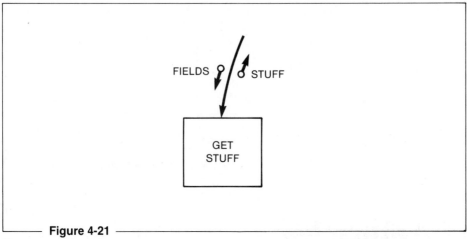

FIELDS   STUFF

GET STUFF

Figure 4-21

Poor parameter names.

FLAGS

We learned already that activity flags are clues to cohesion problems. Directive flags and reporting flags can be clues to other design problems. Let's look at each one.

COUPLING CLUE 5:   Directive flag going up.

PROBLEM: Inversion of authority, and the boss is not a black box.

SOLUTION: Rename flag, or factor special activity out of the boss.

Figure 4-22 shows a worker that passes a flag up to its boss. The worker module, PRINT WITHDRAWAL NOTICE, compares the withdrawal amount to

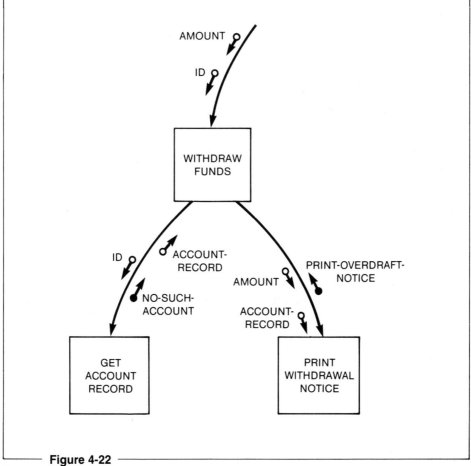

Figure 4-22
Directive flag (needs improvement).

the customer's account balance. If there is not enough money to cover the withdrawal, it sets PRINT OVERDRAFT NOTICE to 'Y' so the boss will print a notice. There are at least two problems here.

1. The boss is not a black box. The worker must know that the boss contains a write statement—that is, that the boss is capable of writing an overdraft notice. The only way one can find that out is by examining the code inside a module.

2. The worker is telling its boss what to do! This is called an inversion of authority and is evidence of poor design. How does the worker know what should be done under these circumstances? The boss might be able to handle overdrafts automatically by transferring funds from another of the customer's accounts; or maybe it extends credit to certain large commercial accounts; or it may print overdraft notices only for amounts exceeding a certain amount. In any case, the worker has no business telling the boss what to do.

How do we solve this design problem? Believe it or not, we simply change the name of the flag to ACCOUNT-IS-OVERDRAWN, turning it into a perfectly acceptable reporting flag. Although only the name is changed, the implications about the flag are eliminated. See Figure 4-23.

Now it is clearly up to the boss to interpret the ACCOUNT-IS-OVERDRAWN flag and decide what to do with an overdrawn account. The worker reports on a situation and "passes the buck" up to the boss where it belongs.

COUPLING CLUE 6:	Reporting flag going up too far.
PROBLEM:	Module that should interpret the flag and do something about it is not doing it; a module at a higher level is handling the flag instead.
SOLUTION:	Move the routine that interprets the flag and then does something about it into the first module that gets the flag and can handle it.

Sometimes the buck gets passed too far in a structure chart, and our clue (once again) is a flag, this time a reporting flag. Figure 4-24 shows a structure chart with a slightly different problem than the one we just studied. In this case, our clue is TRANS-OK, a perfectly acceptable reporting flag. VALIDATE TRANSACTION sets it to 'Y' if the transaction record is valid, and to 'N' if the transaction record is not valid. It sends that little bit of information up to its boss, GET NEXT VALID TRANSACTION, leaving it up to the boss to decide what to do with it. So far so good.

The problem is that GET NEXT VALID TRANSACTION doesn't use the information! It simply passes it up to its boss, UPDATE FILE, which then decides whether to update a record (TRANS-OK = 'Y') or to reject the transaction record it just received by writing it out on a report (TRANS-OK = 'N'). The activity of rejecting the invalid transaction record belongs, not in the boss module, but at the level where the invalid transaction record was first discovered. By removing that activity from the boss, we can avoid passing a flag up to UPDATE FILE. Figure 4-25 shows an improvement.

Now the module GET NEXT VALID TRANSACTION does everything that getting a valid transaction involves—including getting a record, validating it, and rejecting it if it is invalid.

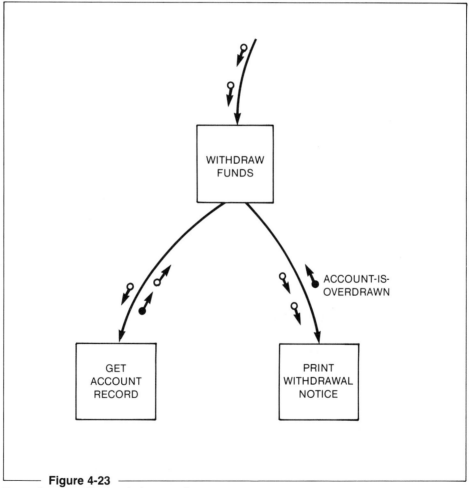

Figure 4-23

Reporting flag: Design is improved.

There are two improvements here.

1. The flag TRANS-OK is isolated to one level in the structure chart, so only two modules care about it.

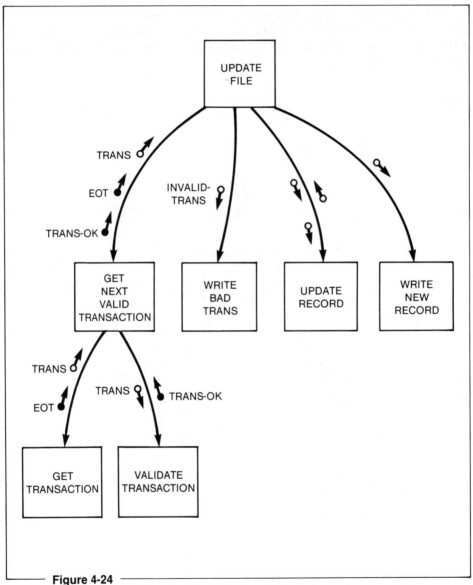

Figure 4-24

Reporting flag (TRANS-OK) travels too far (needs improvement).

2. The boss module at the top, UPDATE FILE, receives only data that is ready for processing; all bad data is rejected at a lower level, and the boss does not have to deal with it.

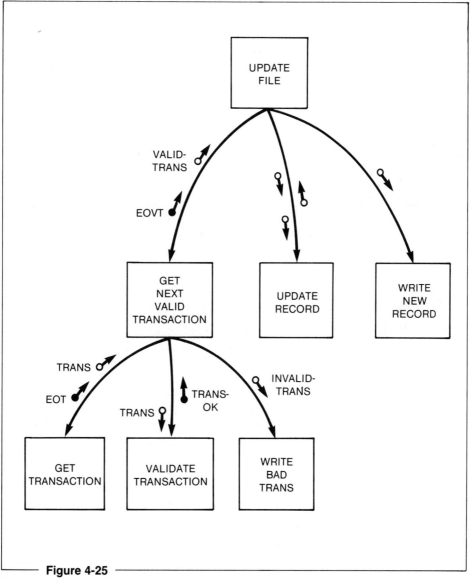

Figure 4-25

Write bad trans module moved to lower level: Design is improved.

COUPLING: A SUMMARY

The study of coupling focuses on the parameters passed between modules. We are concerned with the number of parameters (pass only as many as a module needs to do its job) and the type of parameters (pass only the data parameters that the module needs; reporting flags are necessary and do not present a problem as long as they are not passed through too many modules before they are interpreted; directive flags are clues to design problems and should be changed to reporting flags).

Poor coupling often is a manifestation of poor cohesion: alleviating a coupling problem can fix a corresponding cohesion problem, and vice versa. As we refine a structure chart by improving its coupling and its cohesion, we get closer to our goal of programs assembled from loosely coupled, highly cohesive modules.

SPAN OF CONTROL

Another design-refinement criterion is called *span of control*. This term refers to the number of workers a boss calls and to the number of bosses that call a worker. It is easy to determine span of control. Simply count the number of workers a boss calls, and count the number of bosses a worker has.

SPAN OF CONTROL CLUE 1:   Boss module calling more than seven workers.

PROBLEM: Boss is too complex.

SOLUTION: Add a level of intermediate modules between the boss and the workers.

Our goal is to control the complexity of a module by limiting the number of modules it can call. Generally, a boss should not call more than seven workers, except at transaction centers (see Chapter 3 on transaction centers). A module that calls more than seven workers must keep track of too many interfaces, leaving plenty of room for errors to sneak in. A large number of interfaces is not a problem for a computer, but it is a problem for the human programmers who write, read, and maintain the module. Figure 4-26 shows a structure chart that illustrates the problem.

The boss module must keep track of 11 different modules. By adding a level of modules that remove some of the complexity from the boss, we improve the design as illustrated in Figure 4-27.

Now the boss interfaces with only five modules. Each of the new modules deals with related data items: one handles COST OF SUPPLIES, one handles COST OF SERVICES, and so forth. The boss module's complexity is reduced, making it easier for programmers to write, test, and maintain it. No module

121

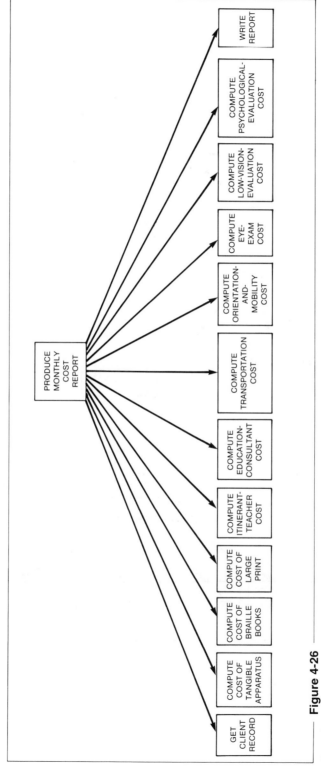

Figure 4-26

Boss has too many workers (needs improvement).

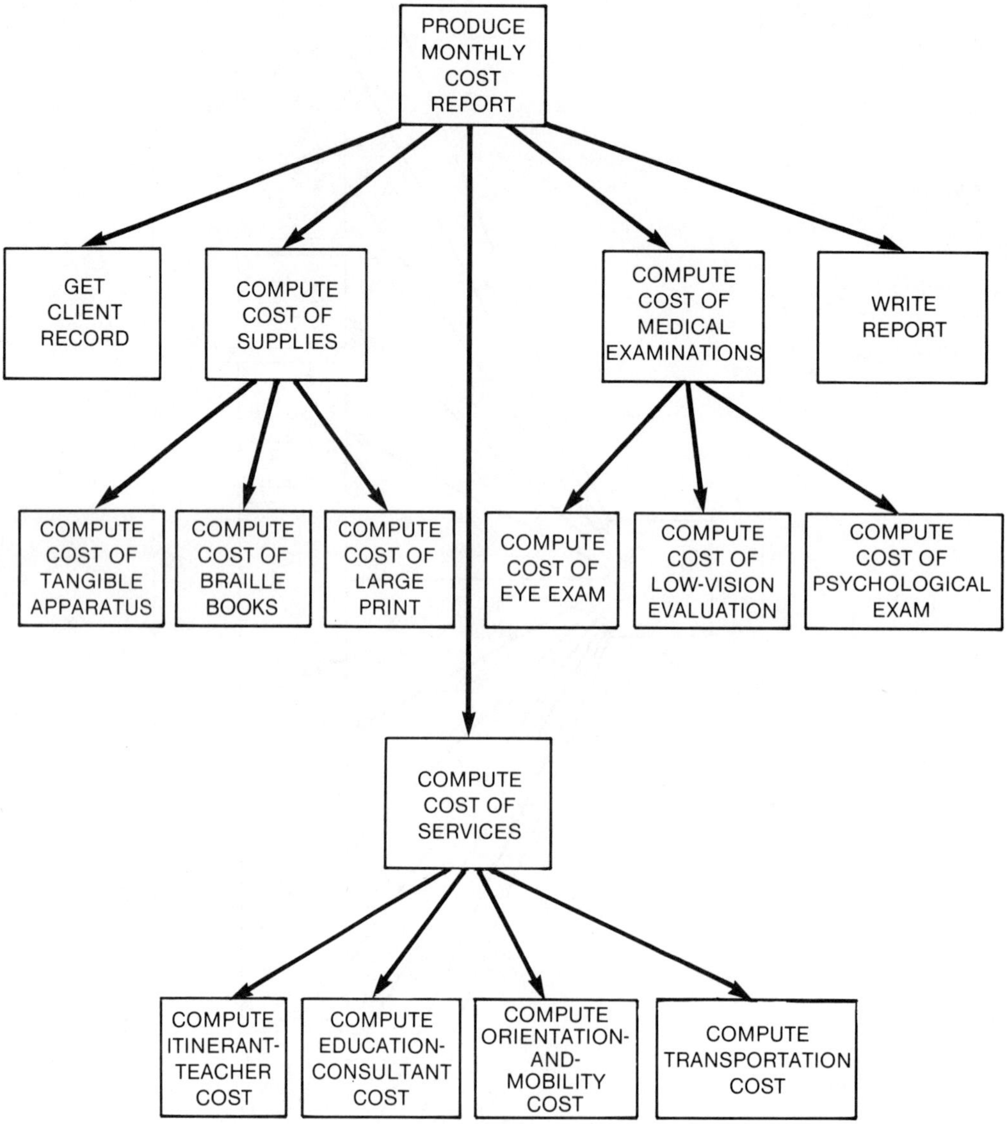

Figure 4-27

Intermediate level of modules: Design is improved.

has to interface with more than five workers. Although the number of modules is greater, the complexity of any single module is reduced.

SPAN OF CONTROL CLUE 2:   Worker is called by only one boss.

PROBLEM: Module is performing a trivial task.

SOLUTION: Include the code for the worker within the boss.

Warning: This refinement activity should take place only after the entire system has been designed and reviewed, and all modules have been specified!

Look at Figure 4-28. The pseudocode for the UPDATE ADDRESS module consists of only one instruction:

```
Move new address to client address
```

Certainly the module serves an important function. After all, changing the client's address is the purpose of the entire structure chart. The other modules

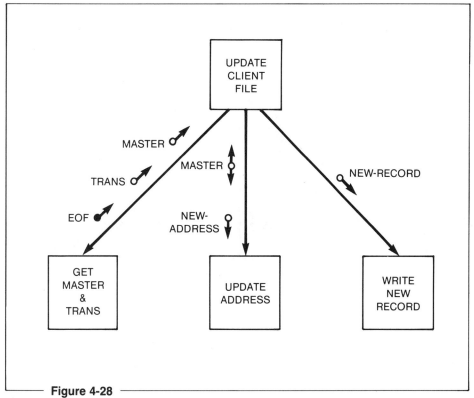

Figure 4-28

Trivial module.

simply get the data in and get rid of it; it is the UPDATE ADDRESS module that is the heart of the program. Still, it is only one instruction. Should it be placed in a module by itself? Perhaps it would be more sensible to incorporate that one line of code directly in the boss module.

We must make a note to the programmer to include the code for the UPDATE ADDRESS module inside the boss, UPDATE CLIENT FILE, rather than making it a called module; we do this by putting a triangle (or "hat") on top of the module as illustrated in Figure 4-29. The structure chart otherwise remains the same. We do not remove the module altogether, because it represents an important function of UPDATE CLIENT FILE. But because of its size, we will actually include it in the boss.

Why do we wait until we have an entire set of structure charts drawn for the programs in the new system before we identify trivial modules? Sometimes a module appears on a given structure chart only once, called by only one boss. It might appear trivial when seen only in the context of one program. But what if that module appears on three or four other structure charts in the set? That indicates that the module is a utility module, one that belongs in a library so other programmers can use it.

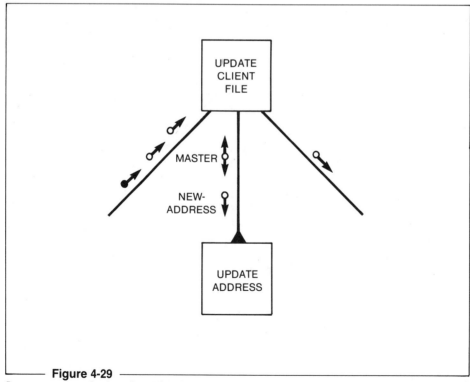

Figure 4-29

Incorporate worker's code within boss.

The problem is that we could decide an otherwise useful module's fate based on incomplete information. So we wait until we have a complete set of structure charts, then find the popular modules—the ones that everybody can use. Those modules ought to be made available to everybody by placing them in a library.

After you draw a complete set of structure charts and specify all of the modules (in Chapter 5 we will learn how to specify modules), go back and study each module that is called by only one boss. *Do not* put a triangle on it—

1. if it could be useful in other applications. For example, a module that computes the number of business days between two dates may be used only once in your system, but it could be useful to other applications in your installation.

2. if it simplifies an already-complex boss. Even if it is only a few lines long, it may be a big help to its boss by performing a function for the boss. Examples include a routine to determine the best retail value of a product, or a routine that crossfoots a matrix.

3. if it is too large to be merged into its boss. If merging the worker into its boss would make the boss too long, don't do it (we'll discuss module size in a later section). Instead, leave the worker in its own module. An example might be a screen-formatting routine that alone requires 45 lines of simple code.

MODULE USEFULNESS

A useful module performs a function that more than one boss wants to invoke. We have seen many examples so far. The beauty of useful modules is that they can be coded once, compiled, and stored in object-code format in a library. They can then be accessed by any programmer who needs them. Maintenance is localized: changes made to a routine are made in one place, not in dozens.

Someday we may assemble programs out of prefabricated modules, just as automobile-assembly plants put together prefabricated parts to make a car. In the meantime, we should identify as many utility modules as possible and exploit them. Some modules that may belong in a module library are

1. file reads, writes, rewrites;

2. application algorithms, such as check-digit calculations and insurance-premium calculations;

3. common business algorithms, such as determining the number of business days between two dates and "julian-to-or-from-gregorian" date conversion;

4. application-data-field validation.

There are literally hundreds of others you can add to this list. Remember that, on a structure chart, vertical stripes down the sides of a module means that it is a utility module in the library.

MODULE SIZE

Although there is no magic number of lines of code in an ideal module, there are some reasonable limits. Try to fit all of the procedural part of a module into one visual picture—that is, on one screen or on one sheet of listing paper.

In installations where programmers work at online terminals, the size of a module should not exceed the size of the screen—which varies, of course. Some screens hold only 10 to 12 lines, but others have a capacity of 20 to 24 lines.

In installations that print source listings on paper, the size of a module should not exceed one page, or approximately 50 to 60 lines.

Although these seem like very strict limitations, remember that these figures are guidelines only. It is simply easier for a programmer (or anyone reading the module) to read and understand a module when not forced to flip pages back and forth, or to scroll instructions on and off the screen.

If you use comments within your program, then do not count them as part of your module. Otherwise you may tend to leave out useful comments simply to produce a "shorter" module. In many cases, comments can be placed together at the beginning of the module's procedural code. If the module is formatted this way, then the executable part of the module remains contiguous, making it easier to read.

Conversely, do not squeeze more than one statement on a line or use abbreviated (and difficult to interpret) data names just so your module remains "within limits." If by keeping a module short enough to fit on a screen or a sheet of paper we make it more difficult to understand and maintain, then we are defeating ourselves.

COMMON STORAGE AREAS

Data fall into two general categories—

data referenced only *within* a module, called *private,* or *local* data;

data accessed by two or more modules, called *shared* data.

Private data are not illustrated on structure charts because they are defined and used only within a module. Shared data, as we have seen, can be illustrated on structure charts as parameters passed between boss and worker modules. However, there is another way to handle shared data—using a *common storage area.*

Using a common storage area results in a program design different from the one we get by passing parameters. There are some important differences between the two methods. In the following discussion we compare them and then suggest some applications in which use of a common storage area could be both appropriate and preferable.

The use of a common storage area reduces the number of parameters passed between modules. Data in a common storage area need not be passed in a parameter list, because the modules that need it already have access to it. As a result, the use of a common storage area eliminates the problem of passing data through modules that *do not* need them, in order to get them to the modules that *do*.

Maintenance on data in a common storage area is localized. This is particularly important if many modules (hundreds, perhaps) access the same data. If the format of a data item passed as a parameter through hundreds of modules changes, then every module it passes through will require updating. However, if the field is defined in a common storage area, its format can be updated in that one place without affecting the modules that access it.

All modules that access data in a common storage area must use the same data names. Because every module uses the same name for the same data, it is easy to scan the modules and find every reference to a common field. In contrast, boss and worker modules need not use the same name for a passed parameter. This flexibility in naming data fields makes writing the code easier (names do not have to match), but it also makes it difficult to identify shared data by scanning the modules.

Data in a common storage area are exposed to all modules having access to the common storage area. Therefore, if data are somehow damaged, all modules that have access to them are suspect. The only way to determine which module is the culprit is to investigate every module. Data passed in a parameter list are less exposed. They are accessible only to those modules that name them in parameter lists.

Thus, there are tradeoffs between using common storage areas and passing shared data in a parameter list. In most cases, passing parameters is preferable to storing data in a common storage area. However, the limited use of common storage areas, coupled with strict control over module access to them, can be beneficial. The key is to restrict access to common storage areas to only those modules that really need it.

APPLICATIONS FOR COMMON STORAGE AREAS

Common storage areas are used both in operating-system software and in application software.

OPERATING SYSTEMS Many operating systems are designed to report automatically on their status in the event of a major system failure. Certain vital information must be available in order to recover and restart the system, information that is changed frequently by many modules. For example, it might be important to know—

the names of all files that were open at the time of the failure;

the last record read from each input file;

the last record written to each output file;

the date and time of the failure;

the name of the program(s) being executed at the time of the failure;

the instruction(s) being executed at the time of the failure;

the current activities being executed at all online terminals.

A great many modules affect such system status information. If every necessary item (only a few examples are listed above) were passed as a parameter in a parameter list, subroutine-call statements would be lengthy, and maintenance could be a nightmare [see Figure 4-30(a)]. Defining those shared data items in a common storage area [see Figure 4-30(b)] gives every module access to them, without passing them from one module to another. As a result, the call statements pass shorter parameter lists, and maintenance on the data in the common storage area is localized.

APPLICATION PROGRAMS   Application programs can also benefit from the judicious use of common storage areas. Whereas in the operating system exam-

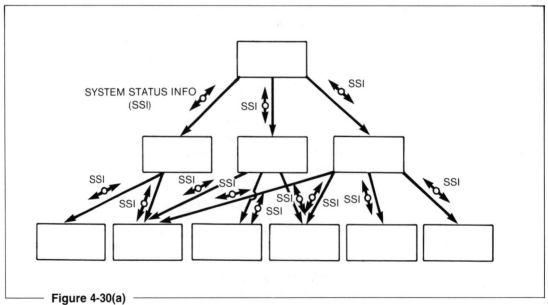

Figure 4-30(a)

System status information passed as a parameter to every module.

ple above the intention was to *allow* access to shared data to all modules, the intention of the following example is to *limit* access to shared data.

Consider the LIBRARY BOOK FILE used in the Agency for the Blind Resource Center. Various modules in the system need access to certain fields in a library book record (Figure 4-31). There are five modules that have access to the LIBRARY BOOK RECORD: RESERVE BOOK, READ LIBRARY FILE, CHECK AVAILABILITY, CHANGE RESERVATION DETAILS, and REWRITE LIBRARY BOOK RECORD. In this example, RESERVE BOOK gets the record from one module only to pass it to another one—it does not use any of the data. CHECK AVAILABILITY needs only CURRENT-STATUS and RESERVATION-DATA. CHANGE RESERVATION DETAILS needs only RESERVATION-DETAILS; and REWRITE LIBRARY-BOOK RECORD needs the entire record. The way it appears in Figure 4-31, however, every module has access to every field.

By storing the LIBRARY-BOOK RECORD in a common storage area, we allow access to the record to each of the four worker modules, while eliminating access to it from the boss [see Figures 4-32(a) and 4-32(b)].

Notice that the number of parameters passed from the boss to each of the workers is reduced because the LIBRARY-BOOK RECORD has been eliminated

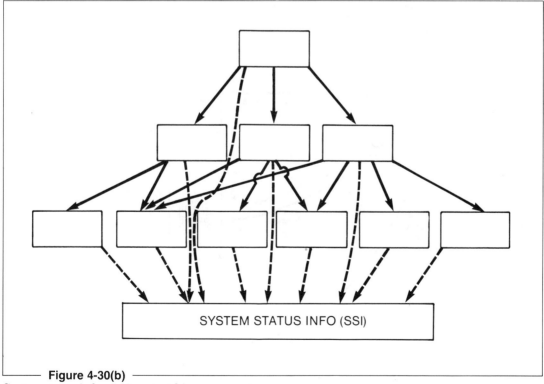

Figure 4-30(b)

System status information stored in common storage area.

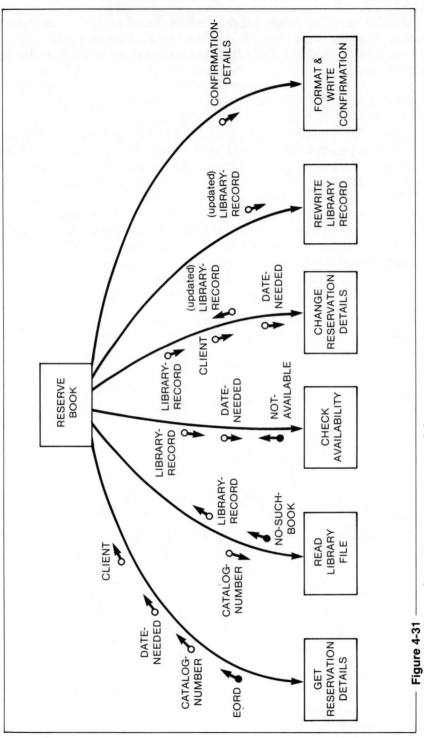

Figure 4-31
Library book record passed as parameter to every module.

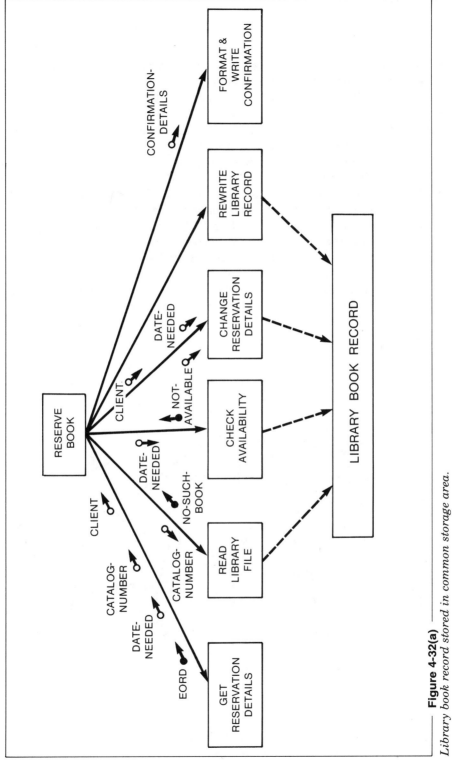

Figure 4-32(a)
Library book record stored in common storage area.

from the list. Designing the program this way has two benefits: (1) It limits access to the LIBRARY-BOOK RECORD to the worker modules, and (2) it localizes maintenance—if the formats of either the LIBRARY BOOK FILE or the LIBRARY-BOOK RECORD ever change, maintenance can be done in one place.

The COBOL code for the modules accessing the common storage area in which LIBRARY-BOOK RECORD is stored appears in Figure 4-33. All four modules are found in the same COBOL program: each is defined with its own entry point (ENTRY) and its own exit point (GOBACK). The DATA DIVISION of a COBOL program is a common storage area, accessible by all modules within the program (within a COBOL program, modules can be defined as *paragraphs* and *sections*). Figures 4-34 and 4-35 illustrate a slightly different structure chart and the COBOL code for the boss module. Notice that the CALL statements to the modules sharing the LIBRARY-BOOK RECORD do *not* include it as a parameter because it is in a common storage area.

Database-management systems (DBMSs) frequently store database records in a common storage area. The modules that are allowed access to the common

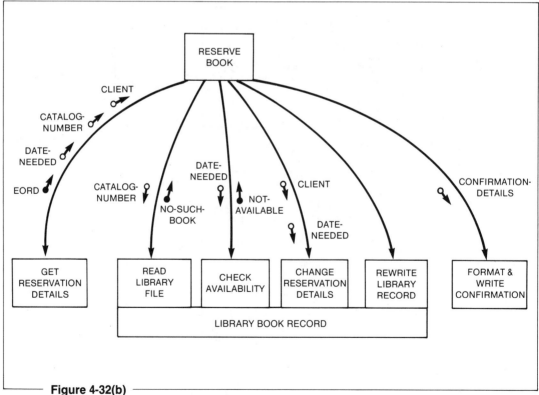

Figure 4-32(b)

Another way to illustrate common storage area.

storage area do so to retrieve or update specific fields for their bosses. This shields the bosses from record details, making both development and maintenance of application programs easier.

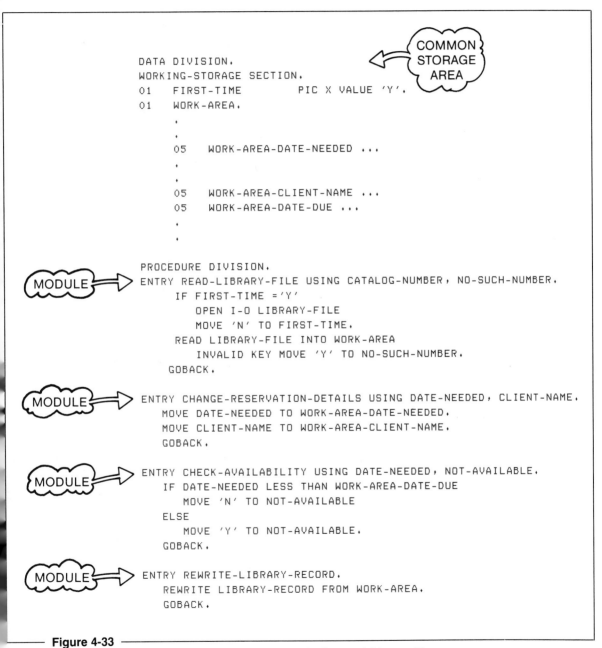

Figure 4-33
COBOL code for modules sharing common storage area in figures 4-32 a and b.

134

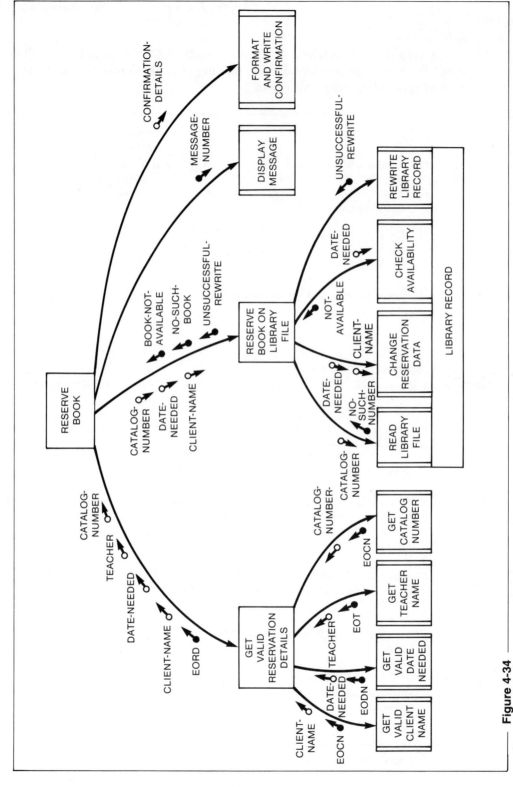

Figure 4-34
Structure chart showing common storage area.

```
PROGRAM-ID.   RESERVE-BOOK.
PROCEDURE DIVISION.
STARTING-POINT.
     PERFORM GET-VALID-RESERVATION-DETAILS.
     IF STOP-FLAG = 'Y'
        GOBACK.

     PERFORM RESERVE-BOOK-ON-LIBRARY-FILE.
        IF BOOK-NOT-AVAILABLE = 'Y'
            MOVE BOOK-UNAVAILABLE-MESSAGE-NUMBER TO MESSAGE-NUMBER
            CALL DISPLAY-MESSAGE USING MESSAGE-NUMBER
            GO TO STARTING-POINT.
        IF NO-SUCH-BOOK = 'Y'
            MOVE NO-SUCH-BOOK-MESSAGE-NUMBER TO MESSAGE-NUMBER
            CALL DISPLAY-MESSAGE USING MESSAGE-NUMBER
            GO TO STARTING-POINT.
        IF UNSUCCESSFUL-REWRITE = 'Y'
            MOVE BAD-REWRITE-MESSAGE-NUMBER TO MESSAGE-NUMBER
            CALL DISPLAY-MESSAGE USING MESSAGE-NUMBER
            GO TO STARTING-POINT.

     CALL FORMAT-AND-WRITE-CONFIRMATION USING
        TEACHER, CATALOG-NUMBER, TITLE, MEDIUM, CLIENT, DATE, RESERVE-FLAG.

     GOBACK.


GET-VALID-RESERVATION-DETAILS.
     CALL GET-VALID-CLIENT-NAME USING NAME, STOP-FLAG.
     CALL GET-VALID-DATE-NEEDED USING DATE-NEEDED, STOP-FLAG.
     CALL GET-TEACHER-NAME USING TEACHER-NAME, STOP-FLAG.
     CALL GET-CATALOG-NUMBER USING CATALOG-NUMBER, STOP-FLAG.

RESERVE-BOOK-ON-LIBRARY-FILE SECTION.
     CALL READ-LIBRARY-FILE USING CATALOG-NUMBER, NO-SUCH-NUMBER.
        IF NO-SUCH-NUMBER = 'Y'
            MOVE 'Y' TO NO-SUCH-BOOK
            GO TO EXIT-RESERVE.
     CALL CHECK-AVAILABILITY USING DATE-NEEDED, NOT-AVAILABLE.
        IF NOT-AVAILABLE = 'Y'
            MOVE 'Y' TO BOOK-NOT-AVAILABLE
            GO TO EXIT-RESERVE.
     CALL CHANGE-RESERVATION-DATA USING DATE-NEEDED, CLIENT-NAME.
     CALL REWRITE-LIBRARY-RECORD USING UNSUCCESSFUL-REWRITE.
        IF UNSUCCESSFUL-REWRITE = 'Y'
            GO TO EXIT-RESERVE.

EXIT-RESERVE.
     EXIT.
```

Figure 4-35

COBOL code for Reserve Book module in Figure 4-34.

SUMMARY

The first-cut structure chart is derived from a dataflow diagram. It usually contains many design flaws that we find by applying design refinement criteria to it. The criteria are cohesion, coupling, span of control, module usefulness, and module size (see Figure 4-36). We also consider using common storage areas to reduce the number of passed parameters.

Cohesion refers to the strength of association of data and instructions inside a module. Ideally, a module completes a single well-defined function each time it is invoked; such modules are highly cohesive. Clues that a module is not cohesive are found in the module's name, the location of the module on the structure chart, and activity flags.

Module name. A module name should have a strong active verb and a singular, specific direct object. Assembly-line names and nonspecific names suggest that a module is performing more than one task. Such modules should be decomposed into several single-function modules.

Location of the module. A module should perform part of its boss's overall function—that is, it should contribute to the completion of its boss's task. A module that does not contribute to its boss's function should be moved so that it reports to an appropriate boss.

Activity flags. An activity flag is a flag that a boss sends down to a worker module to tell it which of its activities to perform. Activity flags indicate that the worker is not a single-function module. If it performed only one

Criterion	Ideal	Examine
Cohesion	Module performs only one function	Module names Activity flags Organization of modules
Coupling	Only data needed by module is passed to it; as few flags as possible	Parameter names Number of parameters Reporting flags Directive flags
Span of control	Boss calls no more than 7 workers Worker called by many bosses	Number of workers Number of bosses calling one worker
Usefulness	Utility modules for library	Module function Parameters
Size	One visual page (screen or hardcopy)	Pseudocode

Figure 4-36

Summary of structure chart refinement criteria.

function, there would be no need for the boss to send down a flag. A module that needs an activity flag should be decomposed into single-function modules. As a result, each of the new modules will probably be much more cohesive, and the flag will be eliminated.

Coupling refers to the dependence one module has on another. Modules that pass parameters are dependent on each other. Our goal is to keep interfacing modules as independent as possible. Coupling problems manifest themselves on a structure chart in both data parameters and flags.

Data parameters.   A module should have access to only the amount of data it needs to do its job—no more than that, and no less. By eliminating unnecessary module access to data, we reduce the exposure that data has. We may also increase the usefulness of a module by building it so that it expects only essential data—rather than, for example, whole data records that contain the needed data. We can sometimes reduce the number of data parameters between modules by collecting related ones into a natural data structure.

Flags.   Flags may or may not indicate design problems. When we studied cohesion, we saw that activity flags are clues to cohesion problems within a worker. On the other hand, reporting flags are ones that worker modules send up to their bosses in order to report on situations that have occurred, such as reaching the end of a stream of input data or detecting invalid data. (Reporting-flag names usually contain adjectives.) Reporting flags are not design problems unless they are passed farther up in a structure chart than is necessary. Conversely, directive flags are flags that a worker returns to its boss telling the boss what to do next. The names of directive flags usually contain imperative verbs. Their names should be changed so that they report on a situation, thus allowing the boss to decide what to do under the circumstances.

Span of control is a measure of the number of workers reporting to a boss, and of the number of bosses a worker has. A boss should interface with no more than seven worker modules; more than seven interfaces introduces complexity, not for the computer but for the human beings who must deal with the code. If a boss calls more than seven workers, introduce a level of modules between the boss and the workers, reducing the number of modules the boss must call.

Modules that are called by more than one boss are good; one of our goals is to identify as many generally useful modules as possible. Modules that are called by only one boss may be performing trivial functions. If the code inside a module will be very brief or very simple, we can include it within its boss in place of the instruction that would call the module. We indicate this by drawing a triangle on the top of the worker module on the structure chart. We do not physically include a module's code within its boss if doing so would make the boss too complex or too long, or if the module has the potential to be a utility module.

Module usefulness refers to the likelihood that a module will be called by multiple bosses. The more generally useful a module is, the more bosses it may have. Utility modules can be coded and compiled, then placed in a library for inclusion by other programs in the installation. Some utility modules that could be found in a library are file-access modules (read, write, rewrite), specific application algorithms (insurance-premium calculations, interest calculations on a loan, computing a student's GPA), common business algorithms (date conversions).

A *module's size* is not as important a design-refinement criterion as the preceding ones, but it should be considered. A module ought to fit on one page— that is, on one visual display (screen sizes range typically from 12 to 24 lines, but sizes vary with manufacturer and applications) or on one sheet of hardcopy (around 50 or 60 lines). This enables a programmer to follow the module's logic without flipping pages or scrolling lines on and off a screen.

Common storage areas can be defined within a program: they allow several modules to access a shared data area. The data stored there need not be passed between modules as part of a parameter list, because the modules that require access to the data have that access. Some advantages of using common storage areas are that maintenance is localized and that the common data are exposed to only a limited number of modules.

KEY WORDS

Activity flag   A flag passed from a boss module to a worker module. The worker module tests the flag and performs a function based on its value.

Black box   Something one can use without knowing how it works.

Cohesion   A measure of the relationship of instructions inside a module. Our goal is high cohesion.

Common storage area   A data-storage area directly accessible by more than one module.

Coupling   A measure of the amount of dependence one module has on another. Our goal is loose coupling.

Directive flag   A flag passed from a worker module to its boss. The name of the flag implies what the boss module's next activity will be.

Local data   Data accessible by only one module; also called *private data*.

Module usefulness   See *utility module*.

Private data   See *local data*.

Reporting flag   A flag passed from a worker module to its boss. The value of the flag indicates the status of some condition; also called a *return code*.

Return code   See *reporting flag*.

Shared data   Data accessed by two modules. They include both passed parameters and data placed in a common storage area.

Span of control   The number of worker modules a boss module calls; also the number of bosses that call a worker module.

Utility module   A module called by more than one boss.

EXERCISES

Use Figure 3-5, the first-cut structure chart for Fill Book Request, to answer the following questions.

1. Factor another module out of GET VALID BOOK REQUEST.

2. Why is FORMAT PACKING SLIP an incomplete name for that module? Rename the module.

3. Does FORMAT PACKING SLIP accomplish one of the major functions of filling a book request? Should FORMAT PACKING SLIP report to FILL BOOK REQUEST? If not, what other module is a more appropriate boss for FORMAT PACKING SLIP?

4. Does FORMAT PACKING SLIP get all the data it needs to produce a packing slip? If not, revise the structure chart so it gets the missing data.

5. (a) Does READ LIBRARY FILE get all the data it needs, and use all the data it gets?
   (b) Does RESERVE BOOK get all the data it needs and use all the data it gets?
   (c) Does CHECK BOOK OUT get all the data it needs and use all the data it gets?
   (d) Revise the structure chart to resolve the discrepancies you identified.

6. What kind of flag is UPDATE-FLAG? Should it be eliminated? If so, revise the structure chart.

7. What kind of flag is NO-SUCH-CLIENT? Should it be eliminated? If so, revise the structure chart.

8. Figure 4-E8 (page 140) shows a first-cut structure chart and supplementary data-dictionary entries for ORDER BOOK. It was derived from bubble 6.3 in the appendix. Produce a refined structure chart for ORDER BOOK.

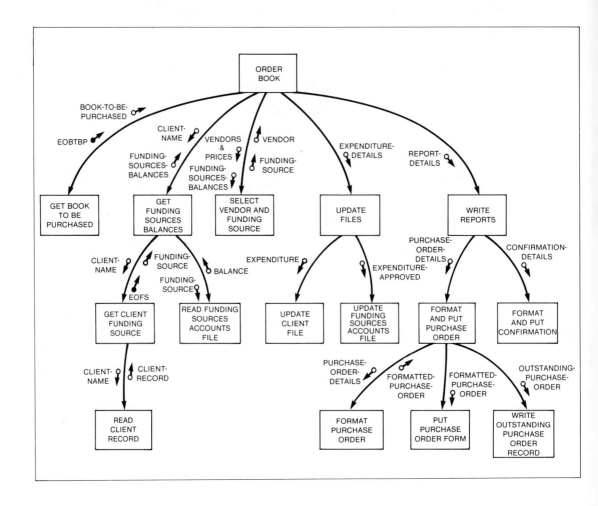

Figure 4-E8

First-cut structure chart and supplementary data dictionary entries for ORDER BOOK.

CHAPTER 5

Specifying Modules

When you complete this chapter, you will be able to—

- identify the appropriate specification techniques for a module based on the module's function
- specify modules using these nonprocedural techniques:

  description of input and output parameters

  algorithms

  card, tape, and disk record layouts

  print report layouts and CRT display layouts

  lists

  decision tables

  decision trees

- specify modules using these procedural techniques:

  flowchart

  pseudocode

  Nassi–Shneiderman chart

INTRODUCTION

Although the structure chart and the data dictionary comprise a large part of the documentation we will pass on to the programmer, our programming specifications are not yet complete. We must now *specify* each of the modules on the structure chart. That is, we must describe the module's *function* in enough detail so that the programmer knows what the module must do when it is executed. Module specifications serve another, more lasting function as well: they are retained as part of the system documentation for reference by other programmers who must understand, change, or replace modules.

In this chapter, we first discuss several techniques for specifying a module *nonprocedurally*. This means that we state clearly and unambiguously what a module is supposed to do, but we do not specify how the programmer should write the program instructions. Next we examine techniques we can use to specify modules *procedurally*. That is, we become more explicit about the way the programmer should implement the module.

NONPROCEDURAL SPECIFICATIONS

There are reasons why we choose to specify some modules nonprocedurally.

1. The function of a module is less likely to change than the implementation of a module. That is, what the module does is less likely to change than how it does it. For example, a module that returns the text of an error message based on an error-message number might at first be coded as a table. As the number of error messages grows, it might be changed to read an indexed sequential file. The function stays the same, though the implementation changes. Therefore, nonprocedural module specifications are likely to prove more lasting documentation than their procedural counterparts.

2. Experienced programmers know the capabilities and shortcomings of their programming languages, often better than the designer (who may not have written any programs recently); they are also familiar with programming techniques that are appropriate for different applications. Thus, experienced programmers are often in a better position to determine the implementation of a module than is the designer.

3. Nonprocedural specifications are less language- or computer-dependent than procedural ones. Thus, the same module specifications can be used by programmers who write different versions of the system in different languages for execution on different equipment. (Consider commercial software written for personal computers, or financial application packages available for different mainframes).

The nonprocedural module-specification techniques we will examine are

1. description of input and output parameters,

2. algorithms,

3. card, tape, and disk record layouts,

4. printed report layouts and CRT display layouts,

5. lists,

6. decision tables,

7. decision trees.

As we discuss each technique, we will identify the kind of modules for which it is an appropriate specification technique, describe the technique itself, and illustrate its use with one or two examples.

DESCRIPTION OF INPUT AND OUTPUT PARAMETERS

This very simple technique is appropriate for modules that perform standard routines—that is, routines that are well known by most programmers. Some examples are modules that read sequential or random files, modules that search tables, and modules that perform standard algorithms such as date conversions.

The technique itself involves stating what the module can expect as input when it is called, and what it is supposed to place in its output parameters upon completion. This includes the values of data parameters and flag settings. It is up to the programmer coding the calling module to satisfy the input-parameter conditions; it is up to the programmer coding this module to satisfy the output-parameter conditions. The programmer is free to code the module any way he wishes (adhering to installation standards, of course), as long as the module performs its task correctly. Let's look at two examples.

EXAMPLE 1: READ MODULE   The module in Figure 5-1 must return the next record in a sequential input file to its boss, along with a reporting flag. Because most programmers are very familiar with reading sequential files, the following nonprocedural module specification is used:

Module name:	READ CLIENT FILE
Input parameters:	none
Output parameters:	
CLIENT-RECORD	If successful read, contains next record; if unsuccessful read, contents unchanged
RETURN-CODE	If successful read, contains '1' If end of file detected, contains '2' If unsuccessful open, contains '3' If unsuccessful for any other reason, contains '4'

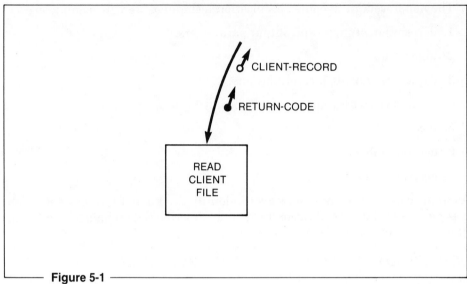

Figure 5-1

Read module.

EXAMPLE 2: DATE-CONVERSION MODULE   Figure 5-2 shows a module as it would appear on a structure chart. Notice that this module receives input parameters, whereas the Read module above did not. By specifying this module nonprocedurally, we assume that the programmer knows how to convert a julian date to its corresponding gregorian date. Here is the nonprocedural specification for this module:

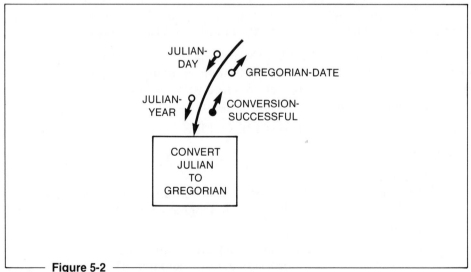

Figure 5-2

Date-Conversion module.

Module name: CONVERT JULIAN TO GREGORIAN

Input parameters:
    JULIAN-DAY   contains integer value 1 through 366
    JULIAN-YEAR   contains integer value 17 through 22

Output parameters:
    GREGORIAN-DATE   date in format MMDDYYYY
    CONVERSION-   contains 'Y' if successful
    SUCCESSFUL   contains 'N' if unsuccessful

In both of these examples, the programmer is free to code the module in any way. As long as the module performs to specifications, it is acceptable.

ALGORITHMS

This specification technique is appropriate for modules that perform calculations—perhaps modules that produce statistics or that perform standard business calculations.

    The technique assumes that the designer and the programmer are familiar with standard algebraic notation. When we use this technique, we simply write the equation, either using parameter names from the structure chart as the operands, or using abbreviations along with a key associating abbreviations with parameter names. Here are two examples.

EXAMPLE 1: CALCULATE REGULAR PAYMENT ON A LOAN   Figure 5-3 shows this module as it would appear on a structure chart. In addition to the

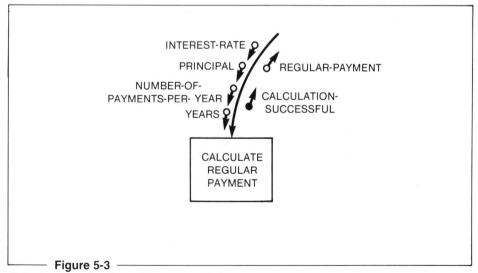

Figure 5-3
Calculate Regular Loan Payment module.

algorithm that follows, we also use the first specification technique to describe how the reporting flag must be set. Notice that we often use more than one technique for a single module. The algorithm part of the module specification is

Module name: CALCULATE REGULAR PAYMENT

$$R = \frac{i*P/N}{1 - (i/N + 1) ** (-N*Y)}$$

where

R is REGULAR-PAYMENT
i is INTEREST-RATE
P is PRINCIPAL
N is NUMBER-OF-PAYMENTS-PER-YEAR
Y is YEARS

EXAMPLE 2: CALCULATE STANDARD DEVIATION   The module that calculates the standard deviation receives a table containing up to 1000 raw test scores, as well as a field indicating the number of scores in the table; it returns the standard deviation, a common statistical value. The module is illustrated in Figure 5-4. The algorithm for the module specification is

Module name: CALCULATE STANDARD DEVIATION

$$SD = \sqrt{\frac{\Sigma\left(\frac{\Sigma X}{N} - X\right)^2}{N}}$$

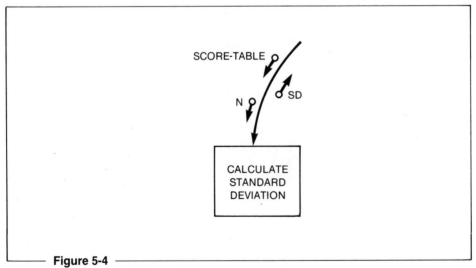

Figure 5-4

Calculate standard deviation module.

where

SD is the standard deviation
X is value of raw test score in table
N is number of test scores

Note: SD is carried to two decimal positions

CARD, TAPE, AND DISK RECORD LAYOUT

This technique is appropriate for modules that perform input/output operations. Those modules are usually found at the lowest level in a structure chart, and they are the ones that need to know the physical details of a file. Therefore, when we specify an I/O module, we must define the file's identification code, its blocking factor, the device on which it resides, its organization, its disposition, and the record descriptions for each record type found in the file.

We can specify I/O modules using a general record-layout form or one designed specifically for the file medium. The goal is to tell the programmer the details of the file the module accesses.

Figure 5-5 shows a part of a structure chart that contains several input/output modules. We will use those modules in the examples that follow.

EXAMPLE 1: FILE DESCRIPTION USING GENERAL RECORD LAYOUT The module called READ CLIENT FILE can be specified using a form like that shown in Figure 5-6. At the top of the form, we describe physical file details. We follow that with a description of the records in the file. Of course, in addition to the file description, we would include in the module specification a statement of how the reporting flag must be set.

One advantage of describing files in this format is that it can be used for all file types stored on any medium having any number of record types of unlimited length.

EXAMPLE 2: CARD-FILE DESCRIPTION The module in Figure 5-5 called WRITE FOLLOW-UP RECORD produces a punched card. A *card layout form* is illustrated in Figure 5-7. At the top of the card layout form we describe general file characteristics. The rest of the form is used to show the location of each field on the punched card. The completed card layout form describing the follow-up record appears in Figure 5-8. In addition to the card layout form, the complete module specification would indicate the conditions under which the module produces each type of output record (notice that there are two different ones).

An advantage of using a card layout rather than the general form illustrated in Figure 5-6 is that a card layout is more graphic. A disadvantage is that it is sometimes difficult to fit a field name into a small space on the form.

REPORT AND CRT DISPLAY LAYOUT

This specification technique is appropriate for modules that format and write output on a printer or on a CRT. We use a two-dimensional grid in order to define the location of all constant and variable information (in contrast to card, tape, and disk layouts, which are essentially one-dimensional). Constants include page and column headers, field identifiers, separators (such as slashes) between

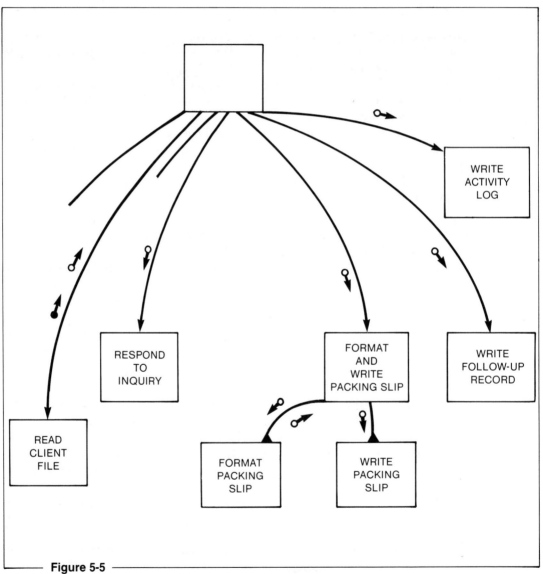

Figure 5-5

Partial structure chart containing input/output modules.

fields or parts of fields, and so forth. Variable data is the actual contents of the report or display—data that is generated by the application program, such as policy numbers, client names, dates of birth, total expenditures, and so forth.

Constants are simply written on the grid in the relative positions at which they will actually appear. The locations of variables are usually indicated with a variety of symbols that tell the programmer the data type and editing considerations. Frequently used symbols and their interpretation appear in Figure 5-9.

The contents of all variable fields must be clearly identified for the programmer. If column headers or field identifiers do not unambiguously and explicitly indicate what data should appear in a location, then appropriate notes to the programmer *must* be included. For example, one column on a report might be called TELEPHONE. Because it is not abundantly clear that the report should contain the client's home telephone number (as opposed to his work number) in that column, a note to that effect must be made on the report layout form. Remember, when we specify modules, our goal is to communicate details to the programmer. If that requires some special comments, then we must include them.

We may also include notes for anything out of the ordinary, such as displaying only the first 10 positions of a 30-character field.

In Chapter 8 we will examine the actual design process for both reports and CRT displays; we will study how to present data in a readable, understandable, nonthreatening way. For now, we want simply to become familiar with the graphics used in report and CRT display formats.

File Name: Agency Client File
File ID: AGCL04
Device: 3330   Volume Serial Number: AG192
Disposition: Keep Expiration Date: 99/365
Blocking Factor: 263 records/block
Record Descriptions
Record Type 1: CLIENT RECORD

Position		Field	Data Type	Storage	
From	To	Name	A, A/N, N	Format	Constant
001	001	COMPLETION-CODE	A		
002	007	CLIENT-ID	N	Z (0 dec)	
008	027	CLIENT-NAME	A/N		
028	028	filler	A		blank
029	033	TOTAL-EXPENDITURES- YTD	N	P (2 dec)	

Figure 5-6

Record layout form for client record.

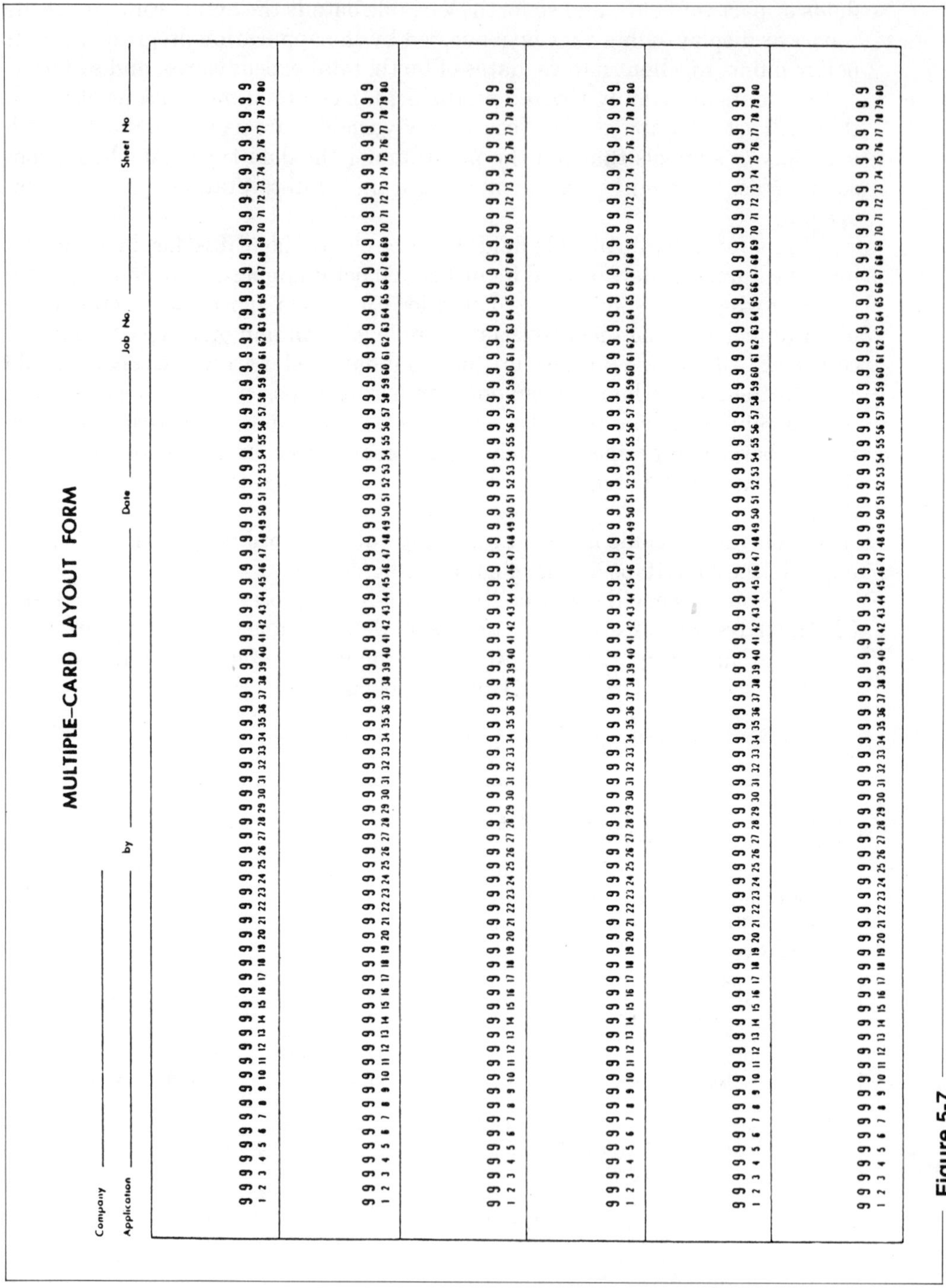

Figure 5-7
Blank card layout form.

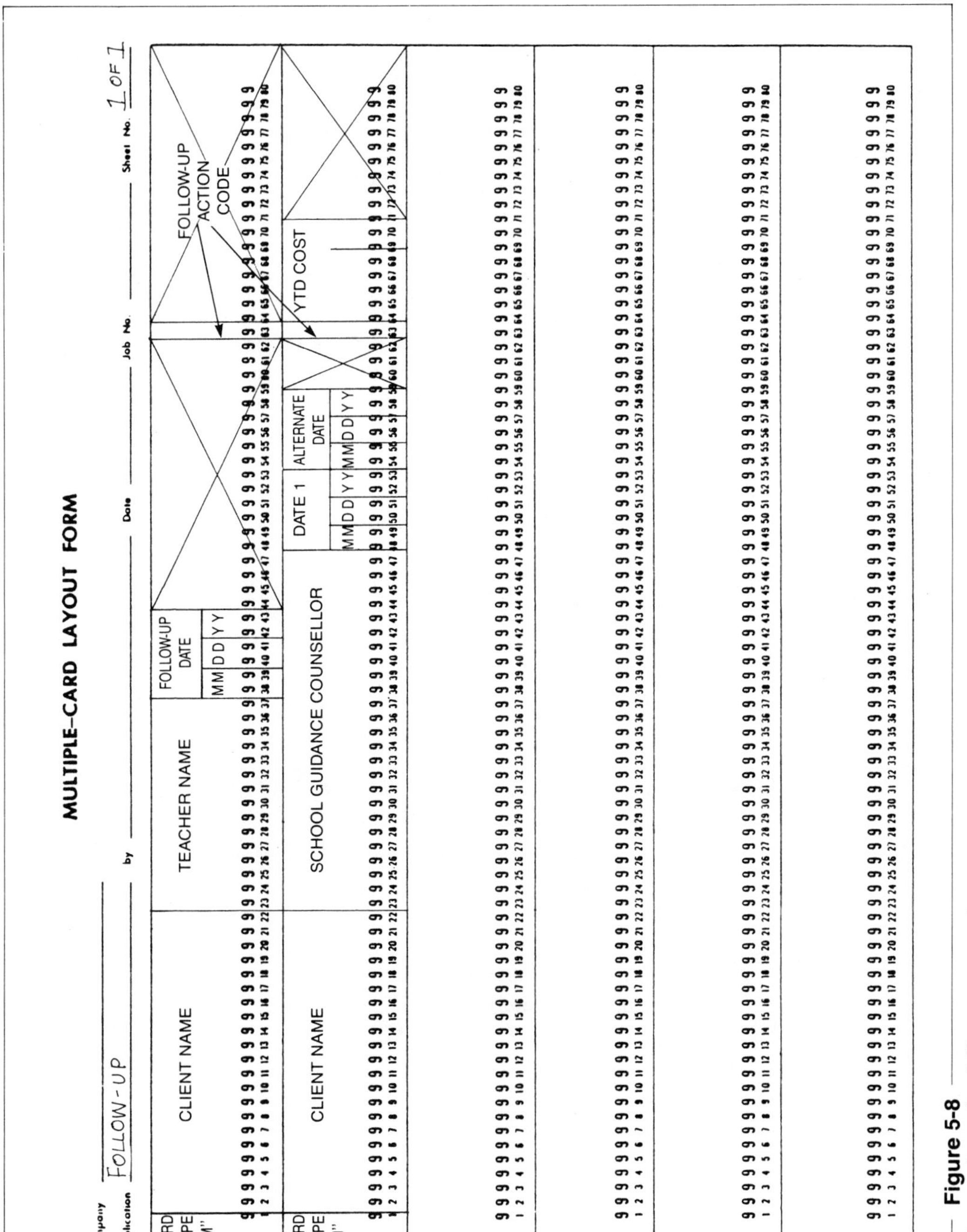

Figure 5-8
Card layout form for follow-up record.

EXAMPLE 1: REPORT LAYOUT   The module in Figure 5-5 called FORMAT AND WRITE PACKING SLIP produces a printed report. The report layout for the packing slip appears in Figure 5-10.

EXAMPLE 2: CRT DISPLAY LAYOUT   The module in Figure 5-5 called RESPOND TO INQUIRY displays data for the CRT operator. The display layout that serves as the module specification is illustrated in Figure 5-11.

LISTS

This module-specification technique is appropriate for modules that process data based on simple conditions. We simply provide the programmer with a list of all conditions that can be encountered and what to do in each case. Examples include modules that validate fields, and modules that return values extracted from tables.

EXAMPLE 1: TABLE OF FIELD-VALIDATION CRITERIA   In Figure 5-12, there is a module called VALIDATE FIELD. The major portion of this module's specification is a table that describes the validation criteria for each of the fields the module validates; the other part of the specification uses the first nonprocedural specification technique to describe how the flag is set. The table appears in Figure 5-13.

The programmer need not code a table in the module. The table is being used only as a nonprocedural specification technique that tells the programmer what the module is supposed to do (not how to do it).

EXAMPLE 2: TABLE OF LONG-DISTANCE TELEPHONE RATES   There is also a module in Figure 5-12 named GET CALL RATE. Based on the time of day, the day of the week, use of operator assistance, and the called country,

SYMBOL	MEANING
X	Alphabetic or alphanumeric character
9	Numeric character
Z	Zero suppress
$	Insert or float $
,	Insert comma
.	Insert decimal point
CR	Display if negative
DB	Display if negative
-	Display if negative

Figure 5-9

Symbols used on report and CRT display layouts.

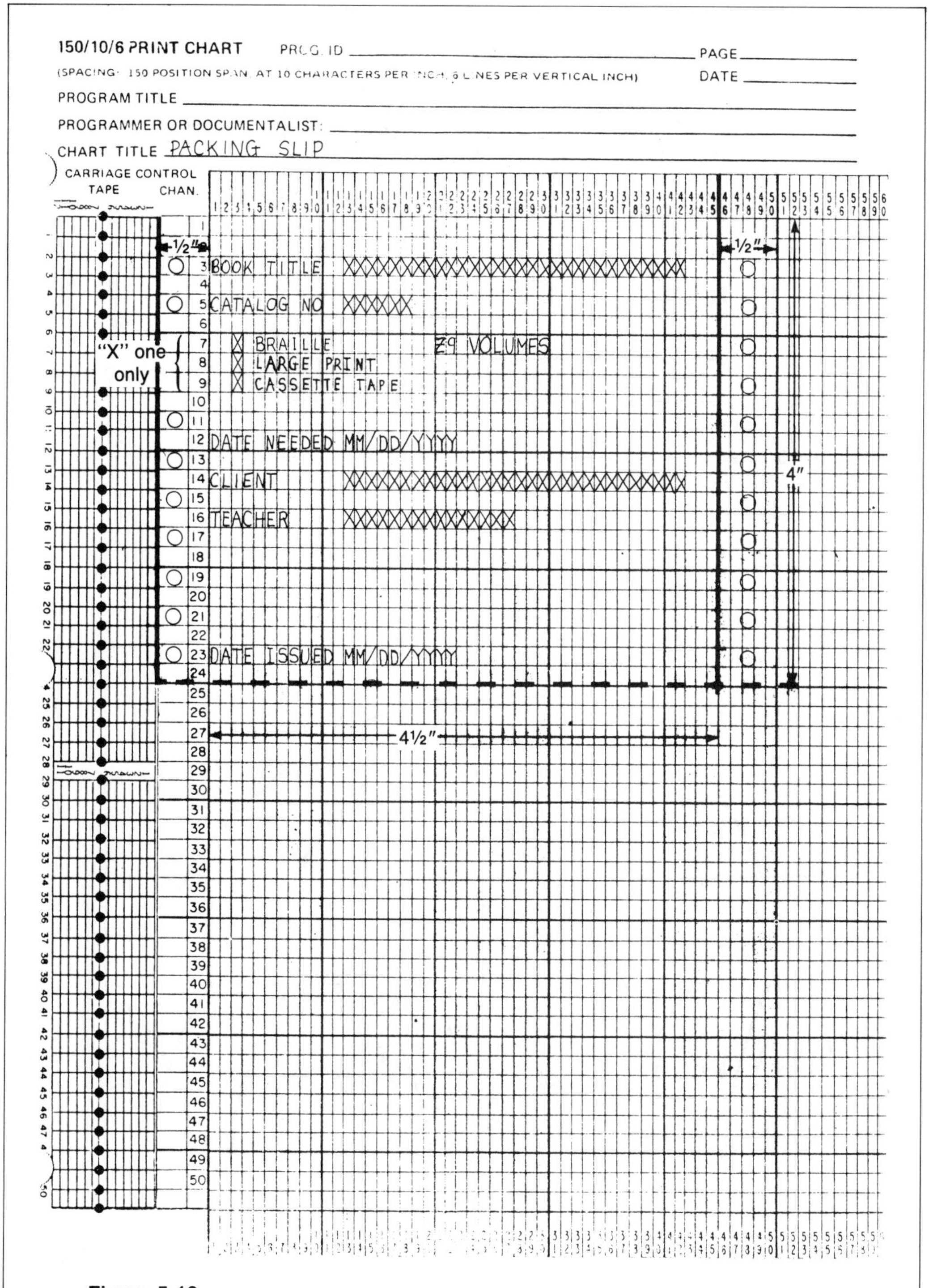

153

150/10/6 PRINT CHART   PROG. ID _____   PAGE_____

(SPACING: 150 POSITION SPAN AT 10 CHARACTERS PER INCH, 6 LINES PER VERTICAL INCH)   DATE _____

PROGRAM TITLE _____

PROGRAMMER OR DOCUMENTALIST: _____

CHART TITLE PACKING SLIP _____

Figure 5-10

Report layout for packing slip.

154

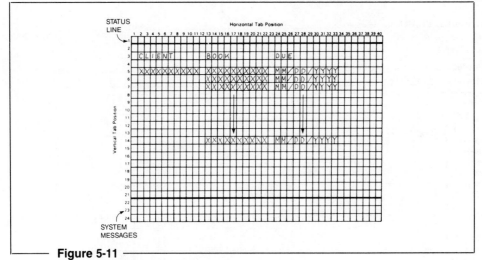

Figure 5-11
Screen layout for RESPOND TO INQUIRY.

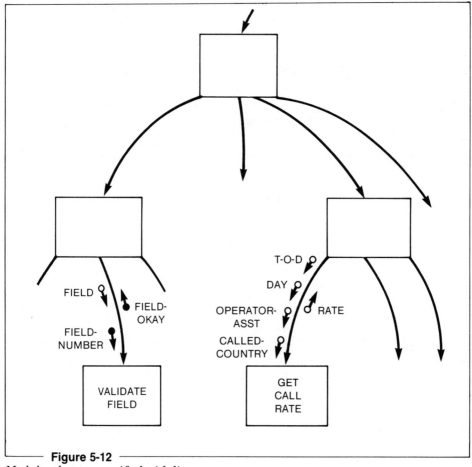

Figure 5-12
Modules that are specified with lists.

FIELD NUMBER	FIELD NAME	VALIDATION CRITERIA
01	Account Number	Numeric Assigned In service
02	Originating Location	Numeric
03	Call Type	1 (dial/station) 2 (operator/station) 3 (operator/person)
04	Destination Location	Numeric
05	Time of Day HH	Numeric, 00–23
06	Time of Day MM	Numeric, 00–59
07	Duration HH	Numeric, 00–99
08	Duration MM	Numeric, 00–59
09	Call Date	Gregorian
10	Calling Card	Numeric Active

Figure 5-13

Field validation criteria.

International Telephone Rates

— Initial 3 Minutes —

	DIAL STATION		OPERATOR ASSISTED STATION		PERSON	
	DAY	NGT/SUN	DAY	NGT/SUN	DAY	NGT/SUN
American Samoa (A)	6.30	5.10	8.40	8.40	12.60	12.60
Andorra, Austria, France, Greece, Liechtenstein, Monaco, Switzerland	6.30	6.30	7.05	7.05	12.60	12.60
Argentina (A), Bolivia (A), Colombia (A) (J), Haiti (E) (J), Paraguay (A)	7.05	5.70	8.40	8.40	12.60	12.60
Australia (A), Brazil (G), China-Taiwan (A), Israel (H), Japan (A), Philippines (A)	7.80	5.85	9.45	7.05	12.60	9.45
Bahrain	9.45	9.45	12.60	12.60	15.75	15.75
Belize (J)	5.70	5.70	6.30	6.30	9.45	9.45
Belgium, Denmark, Finland, German Dem. Rep. (B), Germany, Fed. Rep. of (B), Italy, Luxembourg, Netherlands, Norway, Portugal, San Marino, Spain, Sweden, Vatican City	6.30	5.10	7.05	5.40	12.60	9.45
Chile (G), Peru (G), Venezuela (E) (J)	7.05	5.70	8.40	6.85	12.60	9.45
Costa Rica (E) (J), El Salvador (E) (J), Guatemala (C) (E) (J)	5.70	4.35	6.30	5.25	9.45	7.90
Cyprus (A), Guam (H), Turkey (A), Yugoslavia (A)	7.05	5.70	9.45	7.05	12.60	9.45
Ecuador (E) (J)	7.05	5.70	8.40	6.85	12.60	12.60
Fiji, Guyana (J), Hong Kong, Kuwait, New Zealand	7.05	7.05	8.40	8.40	12.60	12.60
French Antilles (J)	6.15	6.15	8.40	8.40	12.60	12.60
Honduras (E) (J), Nicaragua (E) (J)	5.70	4.35	6.30	6.30	9.45	9.45
Indonesia, Korea, Rep. of, Malaysia, New Caledonia, Papua New Guinea, Singapore, United Arab Emirates	7.80	7.80	9.45	9.45	14.20	14.20
Iran, Iraq, Saudi Arabia, South Africa, Tahiti, Thailand	7.80	7.80	9.45	9.45	12.60	12.60
Ireland, United Kingdom	4.80	3.75	5.70	4.25	10.10	7.50
Kenya	7.05	7.05	8.50	8.50	14.20	14.20
Liberia (A)	7.05	5.70	9.45	9.45	14.20	14.20
Netherlands Antilles (C) (F) (J)	6.15	4.95	8.40	8.40	12.60	12.60
Panama (C) (D) (J)	4.80	3.75	6.30	6.30	9.45	9.45
Romania	7.05	7.05	9.45	9.45	12.60	12.60

All rates shown are for the initial three minutes. The charge for each additional minute is 1/3 the initial three minute dial station rate. Federal excise tax is added on all calls billed in the United States.

Rates are categorized as "Day" and "Ngt/Sun." The "Day" rate is in effect from 5 AM-5 PM, the "night" rate from 5 PM-5 AM unless noted. The Sunday rate applies all day Sunday.

Special Conditions
(A) Reduced rates apply Sunday only
(B) Reduced rates apply nights only
(C) Ngt/Sun. rates also apply on Saturday
(D) Night rate 5 PM-7 AM
(E) Night rate 6 PM-5 AM
(F) Night rate 6 PM-6 AM
(G) Night rate 8 PM-5 AM
(H) Reduced rates apply Sat. and Sun. only. There are no reduced night rates.
(J) Listed is the maximum rate for initial 3 minutes; in many areas the rate is even lower.

All rates are effective as of June 6, 1980.

Figure 5-14

Table of International Telephone Rates.

the module returns the rate of the telephone call. We use the table in Figure 5-14 to tell the programmer what the end results must be. It is not our concern whether or not the programmer decides to implement this module by using a table of long-distance rates. As long as the module returns the correct rate each time it is called, it is acceptable.

DECISION TABLES

Decision tables are appropriate for modules that test combinations of conditions for which there are several processing options. Drawing a decision table enables us to identify every possible combination of conditions, and to state explicitly and unambiguously the processing requirements for each combination. Before we examine the steps we take to draw a decision table, let's look at a finished one (Figure 5-15). This decision table describes a module that produces orders for braille materials. The module is called ORDER BRAILLE BOOK, and it appears in Figure 5-16.

The decision table is divided horizontally into two parts: the top half describes all combinations of the conditions being tested; the bottom half defines the actions that are to be executed for each combination. Each of the numbered columns represents one *rule,* or combination of conditions. A decision table is read one rule at a time. All of the conditions listed in a column are strung together with ANDs; the Xs in the bottom half indicate the processing requirements, or *actions,* for that combination of conditions. For example, Rule 1 in the decision table is read this way:

IF Eligible for APH funds
    AND Book is available from APH
    AND Book is available elsewhere

THEN Write APH order form.

Rule 7 is read:

IF Eligible for APH funds
    AND Book NOT available from APH
    AND Book NOT available elsewhere

THEN Write agency purchase order,
    Write BTA braille request.

What action is taken if the client is not eligible for APH funding, the book is available from APH, and the book is not available elsewhere? If you said "Write APH order form, Write agency purchase order," you are correct. Rule 6 contains that combination of conditions.

We will illustrate the steps involved in building a decision table using the module called ORDER BRAILLE BOOK that appears in Figure 5-16. Here is a narrative description of the module's function.

		Rules							
		1	2	3	4	5	6	7	8
Conditions	Eligible for APH funds?	Y	N	Y	N	Y	N	Y	N
	Book available from APH?	Y	Y	N	N	Y	Y	N	N
	Book available elsewhere?	Y	Y	Y	Y	N	N	N	N
Actions	Write APH order form	X	X			X	X		
	Write agency purchase order		X	X	X		X	X	X
	Write BTA order form			X	X				
	Write BTA braille request							X	X

Figure 5-15

Decision table.

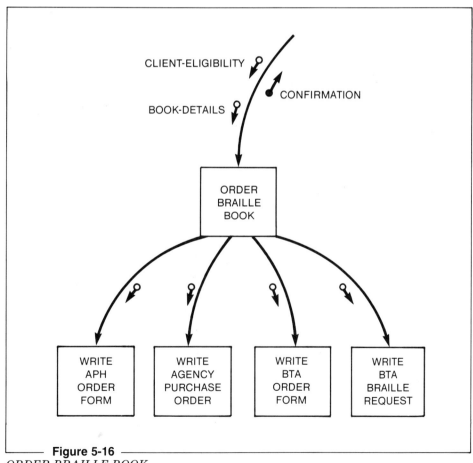

Figure 5-16

ORDER BRAILLE BOOK.

ORDER BRAILLE BOOK

If the child receives APH quota funds, then order the book from APH. You don't need an agency purchase order (P.O.), because no money changes hands. If the book isn't available from APH you have to check the general catalog to see if it's available from someone else. If it is available from someone else, buy it from them. Fill out an agency P.O. and a Braille Transcription Association (BTA) order form; BTA takes care of the details of ordering a copy of the book. On the other hand, if the book is not available anywhere, have it brailled by BTA. For this you need an agency P.O. and a BTA braille request. If the child is not eligible for APH quota funds, but APH has a copy of the book, buy it from APH: fill out a copy of the agency P.O. and a copy of the APH order form.

Using the narrative as our source of information, we will draw a decision table that more clearly describes what actions to take under what conditions.

Step 1.   *Determine number of conditions.*
In this case there are three. (1) Is the client eligible for APH quota funds? (2) Is the book available through APH? (3) Is the book available through some other organization?

Step 2.   *Determine number of condition combinations, or rules.*
We derive this number by multiplying together the number of possible answers for each condition. The first condition (is the child eligible for APH quota funds?) has two possible answers, yes and no. The next condition (is the book available through APH?) also has two possible answers, yes and no. The third question (is the book available through some other organization?) has two answers, yes and no. The total number of combinations is $2 \times 2 \times 2$, or eight.

Step 3.   *List all the conditions* on a piece of paper, one on top of the other:
Eligible for APH funds?
Book available from APH?
Book available elsewhere?

Step 4.   *Draw one column for each rule.*
In this case there are eight rules. See Figure 5-17.

Step 5.   *Fill in all combinations of conditions.*
Start either with the top or the bottom row: we happened to start at the bottom. The number of rules is 8; the number of possible answers to the condition (is book available elsewhere?) is 2. Therefore, 4 of the rules ($8/2 = 4$) will contain the answer Yes and 4 will contain the answer No. Fill in 4 Ys, then 4 Ns. Move up to the next condition (book available from APH?). There are 2 outcomes to this condition. For each group in the row beneath it (4 in a group), 2 of them will be Yes and 2 will be No ($4/2 = 2$). Fill in 2 Ys, then 2 Ns, then 2 Ys, and so forth, until you have filled the entire row. In the top row representing the final condition (eligible for APH funds?),

there are 2 possible outcomes, yes and no. For each of the groups in the row beneath this one (2 per group), 1 will be Yes and 1 will be No (2/2 = 1). Fill in 1 Y followed by 1 N followed by 1 Y and so forth until the table is filled. See Figure 5-18.

Step 6. *Write all possible actions* in the bottom half of the decision table. There are four actions: write an APH order form, write an agency purchase order form, write a BTA order form, write a BTA braille request. See Figure 5-19.

		Rules							
		1	2	3	4	5	6	7	8
Conditions	Eligible for APH funds?								
	Book available from APH?								
	Book available elsewhere?								

Figure 5-17
Columns for decision table rules.

		Rules							
		1	2	3	4	5	6	7	8
Conditions	Eligible for APH funds?	Y	N	Y	N	Y	N	Y	N
	Book available from APH?	Y	Y	N	N	Y	Y	N	N
	Book available elsewhere?	Y	Y	Y	Y	N	N	N	N

Figure 5-18
Decision table rules filled in.

		Rules							
		1	2	3	4	5	6	7	8
Conditions	Eligible for APH funds?	Y	N	Y	N	Y	N	Y	N
	Book available from APH?	Y	Y	N	N	Y	Y	N	N
	Book available elsewhere?	Y	Y	Y	Y	N	N	N	N
Actions	Write APH order form								
	Write agency purchase order								
	Write BTA order form								
	Write BTA braille request								

Figure 5-19
Decision table with actions.

Step 7. *Indicate which action(s) should be taken for each rule.*
This requires going back to the narrative and studying it carefully. As you find a combination of conditions in the narrative, locate that rule in the decision table and indicate the appropriate action. Sometimes a statement that appears later in a narrative will supercede a statement made earlier. When this happens simply revise the action for that rule. You should now have the complete table shown in Figure 5-15.

Step 8. *Check any ambiguities* with the user, the analyst, or someone else in authority. It is *always* wise to check out the *entire* decision table with someone in authority, but it is particularly important if you have encountered any apparent ambiguities or holes after drawing the decision table. If you discover one and do not resolve it, then the error will become part of a program.

Remember that, when we specify a module with a decision table, we are telling the programmer precisely what to do under every possible set of circumstances. We must be sure that *we* know what to do first.

DECISION TREES

Like decision tables, decision trees are useful and appropriate for specifying modules that handle compound conditions. Although decision tables and decision trees can be used interchangeably, decision tables seem more appropriate when the number of actions is small (there are fewer actions than rules) and several rules result in the same action, or when several actions are taken for each rule. Decision trees seem more appropriate when each rule requires different processing (the number of actions is almost the same as the number of rules), and when the test of one condition is based on the outcome of a previous one.

Consider the insurance application module in Figure 5-20. The Insurance Risk Code is one factor used to determine someone's insurance rate. Its value is based on the region in which the insured lives, his or her age, and the last time he or she had an accident. The following narrative describing the module's function was developed after interviewing the user. For simplicity, masculine pronouns are used to refer to the insured, regardless of sex.

ASSIGNING INSURANCE RISK CODE

If the insured lives in Region A, then if he is under 21, his risk code is 104 if he has had an accident in the last year, 117 if his last accident occurred between one and two years ago, and 259 if it happened between two and five years ago; if he has never had an accident, or if his latest one occurred over five years ago, then his risk code is 257. If the insured is 21 years old or older, then his code is 257 if he has never had an accident, 136 if he had an accident in the past year, and 249 if his latest one happened between one and three years ago. Figuring the risk code in Region B is a little different, because there are more age ranges

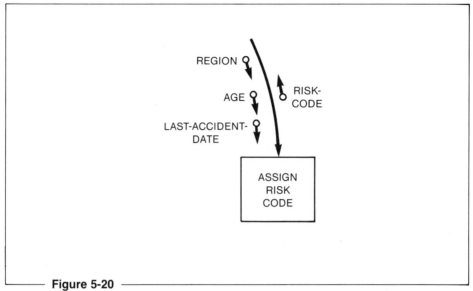

Figure 5-20
Assign Insurance Risk Code module.

to consider: 16 through 18 years old, over 18 through 21, and over 21. For a 16–18-year-old, his code is 072 if he has had an accident in the past six months, 165 if he has had one between six months and a year ago, and 188 if his latest accident occurred between one and two years ago. If he never had an accident, his code is 257. The code is also 257 for an 18–21-year-old who has never had an accident, or if his latest one occurred more than five years ago; it is 118 if his latest one occurred in the past 18 months, and 136 if it happened between 18 months and five years ago. If he's over 21, then the code is 218 if he had an accident in the past two years, and 257 if he did not.

The conditions being tested are region, age, and date of last accident. Region is the *major* condition, because it is dependent on no other condition. Age, on the other hand, is an *intermediate* condition, because it is subordinate to Region: the possible values of age depend on the region in which the insured lives. Finally, Date of Last Accident is the *minor* condition because it depends on Age. The decision tree for the module ASSIGN INSURANCE RISK CODE appears in Figure 5-21.

Reading a decision tree is like reading a flowchart sideways (rotate the figure 90° clockwise, and it looks even more like a flowchart). Notice how many age ranges there are, and how many accident ranges within age groups. Also notice the variety of insurance risk codes; there is almost one code for each combination of conditions.

A decision tree has two parts: the *rules* and the *actions* (much like a decision table). Within the rules area, which appears on the left side of the tree, the major test condition appears at the far left (Region?), the minor condition at the far right (When did last accident occur?), and intermediate conditions in

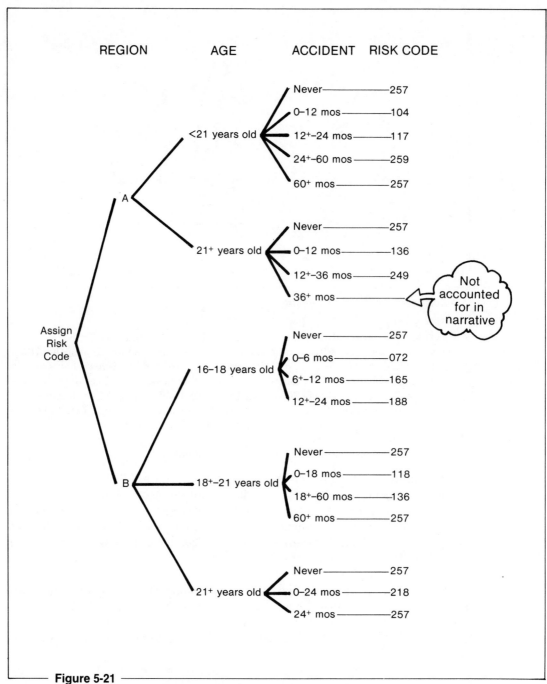

Figure 5-21

Decision tree for ASSIGN INSURANCE RISK CODE.

between (Age?). The far right column indicates the action taken for each rule (Risk Code assigned).

Drawing a decision tree requires some amount of intuition and some amount of trial and error. As we develop a tree, we sometimes make false starts. (Such as choosing the wrong condition as the major one. Draw the decision tree starting with Age as the major condition, and you will soon encounter problems). Don't worry about false starts. Paper is cheap, so throw it away and start over.

Determining the major, intermediate, and minor conditions means reading the narrative through several times and making notes on it. Many narratives describing compound conditions are not even as clear as the one presented here. If you do not understand something, get clarification from the user or from someone else who understands the process being described. Then draw the tree using Figure 5-21 as a model. Starting at the left, write the name of the condition being tested at the top of the page. Then in the column underneath the condition, draw and label one branch for each possible outcome of that condition. Write the name of the next condition above the next column, and draw and label one branch for each possible outcome of this condition *for each of the branches in the previous column.* (For example, age range depends on the region. We drew one set of branches for age range for Region A and one set for Region B). Continue this process until you have drawn all the branches for all the outcomes of all the conditions.

The last step is to fill in the action column, which appears at the far right of the decision tree. You should read the narrative several times, filling in actions as you encounter rules. If you discover ambiguities or holes in the narrative (it is not unusual), then do not make any assumptions. Always clarify the decision tree with someone in authority. In fact, it is a good idea to confirm the entire decision tree with the user, even if you did not encounter any problems.

Remember, the programmer will translate this module specification into code. Any errors that go undetected in the decision tree will become processing errors in the system later on.

PROCEDURAL SPECIFICATIONS

In this section, we examine three techniques with which we can specify modules procedurally. The techniques are drawing *flowcharts,* writing *pseudocode,* and drawing *Nassi–Shneiderman charts.* In specifying a module procedurally, we focus on *how* it carries out its job. This is in contrast to nonprocedural specification techniques, which focus on *what* the module is supposed to do, leaving implementation decisions up to the programmer. When we use flowcharts, pseudocode, or Nassi–Shneiderman charts, we describe the sequence in which the module's instructions will be written. The following are some advantages of specifying modules procedurally.

1. There is little room for interpretation (or misinterpretation) by the programmer. The basic logic of the module is defined, and the programmer simply follows it.

2. Module logic can be reviewed before any code is written. This helps to reveal errors that, if they go undetected, become errors in the module itself. The sooner errors are detected, the easier it is to correct them.

3. Specifying modules procedurally helps novice programmers and programmers who are unfamiliar with the system being developed.

Therefore it is *especially* useful to specify controlling modules procedurally. They are usually found near the top of a structure chart, and they often contain fairly complex decisions that control the invoking of the subordinate modules on the structure chart. The more complex the logic in such modules, the more sense it makes to specify them procedurally. It is important to determine precisely *how* controlling modules do their jobs before writing any code; then the logic can be reviewed (see advantage 2 above) and fixed if there are any errors, and the programmer can write code according to the already-reviewed logic in the specification (see advantage 1 above).

THE BASIC CONSTRUCTS

This text assumes that the reader already understands programming logic. By way of review, this section presents the three basic controlling constructs from which programs are built. Then we can describe how each of the constructs is illustrated by the procedural specification techniques. The basic constructs are

1. sequence,

2. repetition,

3. decision.

Sequence refers to the situation in which control passes from one instruction to the next, so that actions are executed one right after another. Unless we tell a computer to do otherwise, it executes instructions in sequence.

Repetition refers to the situation in which control remains in a loop until some terminating condition is met. Thus, an instruction group consisting of one or several instructions can be repeated a number of times. We must identify (1) the instruction group to be repeated and (2) the terminating condition.

A *decision* is a point at which, out of a set of mutually exclusive instruction groups, one instruction group is selected for execution based on the results of some test condition. We must identify (1) the test condition and (2) the instruction group to be executed for each of the possible outcomes of the test.

Each of the basic constructs can be *nested*. This means that any instruction group can contain sequences, decisions, repetitions, and combinations of these.

For example, one of the actions executed as the result of a test condition might be another decision. We might make a decision inside a repetition; or we might execute one loop inside another one.

When we specify a module procedurally, we specify instruction sequences, decisions, and repetitions. The programmer must then write code to match the basic module logic we have defined. Now we will examine the three alternatives we can use to illustrate the logic inside a module: flowcharts, pseudocode, and Nassi–Shneiderman charts.

FLOWCHARTS

A flowchart is a diagram consisting of symbols that represent actions and decisions, and arrows connecting the symbols to indicate flow of control from one action or decision to another. Words are written inside each symbol to describe the specific action or decision. Flowcharts have been used in data processing for a very long time, and a wide variety of symbols has evolved, each symbol representing a certain type of activity. We will limit our discussion to those symbols that are needed to illustrate the basic constructs of sequence, decision, and repetition.

A rectangle (□) indicates an instruction group. Rectangles connected by arrows indicate sequence: after the instruction group in one rectangle has been executed, the instruction group in the second rectangle is executed. The direction of the arrow shows the flow of control from one instruction group to another. Sometimes the arrowhead is omitted—the flowchart is then read from the top down, or left to right.

A diamond (◇) indicates that a *decision* must be made to follow one control path or another, based on the results of the test condition written inside the diamond. The arrows exiting from the diamond indicate the various paths that can be followed; each path is clearly identified. All the control paths rejoin at a common point (see Figure 5-22).

Repetition is illustrated by combining symbols: the terminating condition is written inside a *diamond,* while the instruction group to be repeated is indicated by one or several *rectangles;* an *arrow* leading back up to the test condition completes the illustration (see Figure 5-23).

Flowcharts allow us to nest one control construct within another by replacing rectangles with decisions and loops. Figure 5-24 shows some examples of nesting as they might appear on flowcharts.

In order both to illustrate and to compare the three procedural specification alternatives, we will present the module specification for a single module using each of the alternatives. The structure chart in Figure 5-25 contains a module called UPDATE MASTER FILE. This is a controlling module, and it is appropriate to specify it procedurally. The module specification for UPDATE MASTER FILE using a flowchart appears in Figure 5-26.

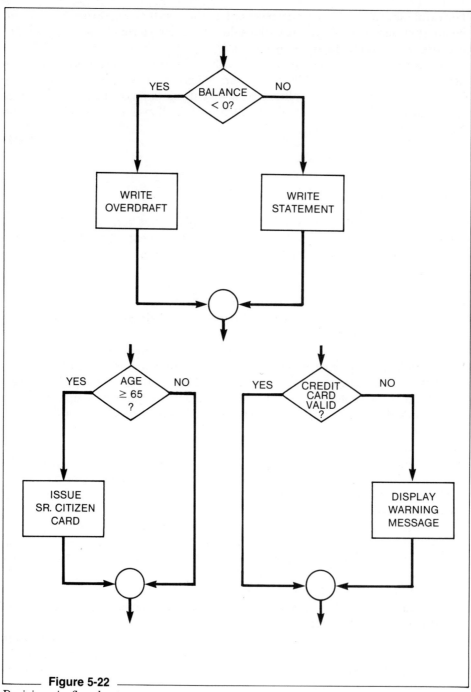

Figure 5-22
Decisions in flowcharts.

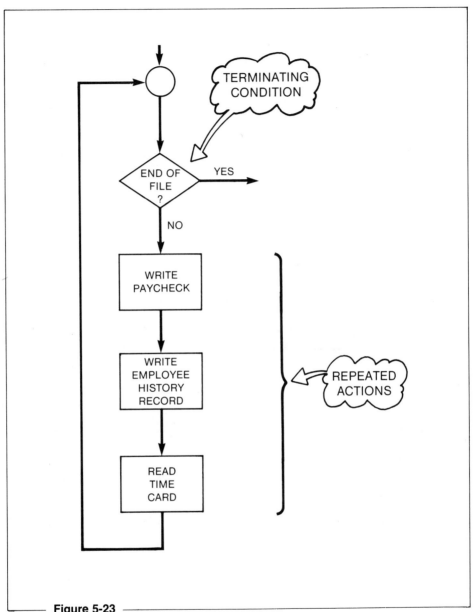

Figure 5-23
Repetition illustrated in flowchart.

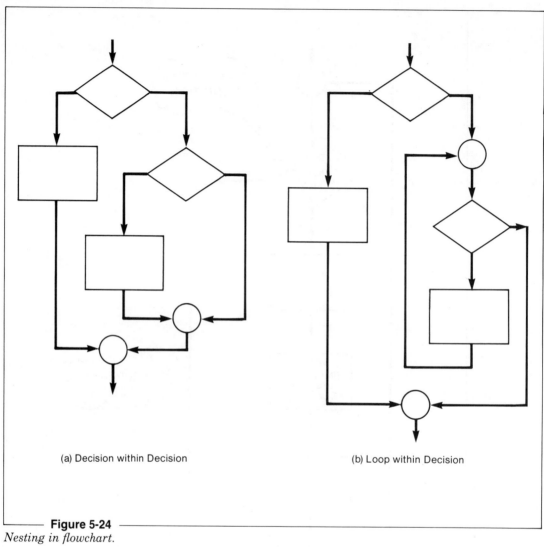

(a) Decision within Decision

(b) Loop within Decision

Figure 5-24
Nesting in flowchart.

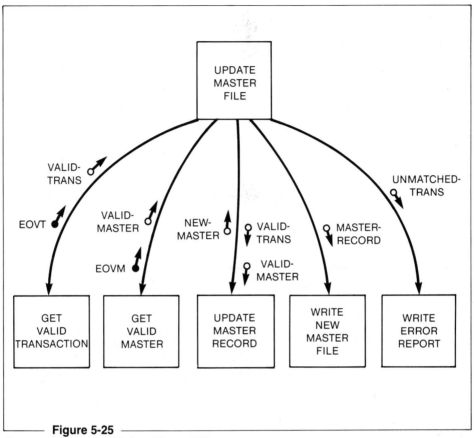

Figure 5-25
Structure chart for Update Master File.

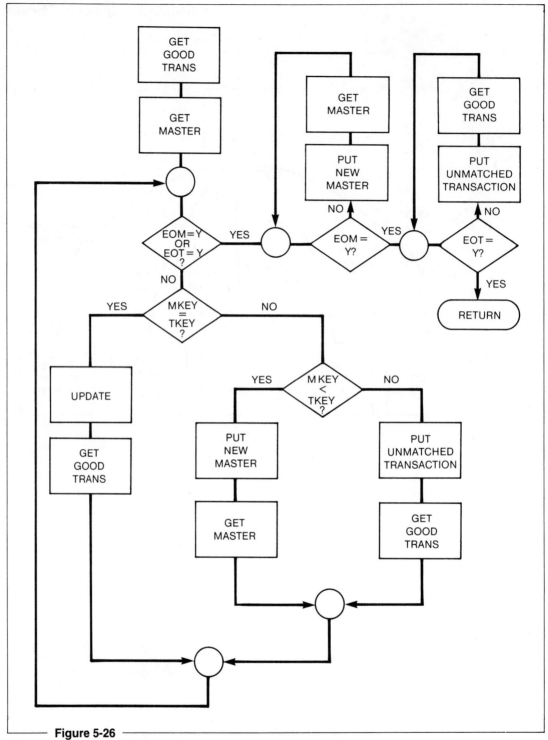

Figure 5-26
Flowchart for Update Master File module.

170

PSEUDOCODE

The prefix *pseudo* comes from a Greek word meaning "fake." Pseudocode is fake code—it looks a lot like program instructions, but it is not the real thing. Pseudocode combines a small vocabulary of keywords with normal English-like sentences, and it uses some special formatting techniques to illustrate sequence, decisions, and repetition.

When we write pseudocode, we write only one statement on a line. *Sequence* is illustrated by writing one line after another.

To illustrate decisions, we use the keywords IF, THEN, ELSE, ENDIF in the following format (note the use of *indentation*):

```
IF condition to be tested
    THEN instruction group executed when condition is true
    ELSE instruction group executed when condition is false

ENDIF
```

To illustrate *repetition* in pseudocode, we use the keywords REPEAT, UNTIL, and ENDREPEAT in this format (note the use of indentation):

```
REPEAT UNTIL terminating condition
    instruction group to be executed as long as the terminating
    condition is false

ENDREPEAT
```

The instruction group inside the loop is executed over and over again, until the terminating condition is satisfied; when that happens, control passes to whatever instructions follow ENDREPEAT. (The keywords REPEAT and ENDREPEAT can be replaced with DO and ENDDO, PERFORM and END-PERFORM, or LOOP and ENDLOOP.)

A variation on the loop in pseudocode uses the keyword WHILE in place of the keyword UNTIL:

```
REPEAT WHILE terminating condition
    instruction group to be executed as long as the terminating
    condition is true

ENDREPEAT
```

You have seen that indentation is used in order to make pseudocode more readable. In order to illustrate *nesting* in pseudocode, we make use of indentation. Figure 5-27 shows some examples of nesting in pseudocode.

The pseudocode specification for UPDATE MASTER FILE is illustrated in Figure 5-28. Compare it to the flowchart in Figure 5-26.

A nested loop

```
    REPEAT UNTIL END-OF-FILE = 'YES'
        WRITE EMPLOYEE-DETAILS
        REPEAT UNTIL NO-MORE-DEPENDENTS
            WRITE DEPENDENT-NAME
            DEPENDENT-COUNTER = DEPENDENT-COUNTER + 1
        ENDREPEAT
        WRITE DEPENDENT-COUNTER
        READ EMPLOYEE-FILE AT END END-OF-FILE = 'YES'
    ENDREPEAT
```

A nested decision

```
    IF CLIENT-STATUS = BLIND
        THEN
            IF AGE LT 3
                THEN DO PRESCHOOL-ROUTINE
                ELSE DO SCHOOLAGE-ROUTINE
            ENDIF
        ELSE
            DO VISUALLY-IMPAIRED-ROUTINE
    ENDIF
```

Figure 5-27

Nesting in pseudocode.

NASSI–SHNEIDERMAN CHARTS

Nassi–Shneiderman charts, named after their developers, resemble pseudo-code in that they use English-like statements to indicate actions. However, like flowcharts, Nassi–Shneiderman charts make use of different shapes to illustrate the three control constructs.

Sequence is illustrated by writing one instruction inside a rectangle and *stacking rectangles* one on top of another: the stack is read from the top down (like flowcharts without any arrows or space between rectangles).

Repetition in a Nassi–Shneiderman chart is illustrated with a symbol that looks like an *upside-down* L (see Figure 5-29). The terminating condition is written inside the L, and the instruction group to be repeated is written underneath (and enclosed by) it.

Decisions in a Nassi–Shneiderman chart are indicated with a *large rectangle divided into three parts* (Figure 5-30). The test condition is written at the top of the rectangle. Underneath the condition, the rectangle is divided vertically in two. One half contains the instruction group executed when the condition is

Update Master File

```
BEGIN Update Master
    CALL Get Good Trans
    CALL Get Master

    REPEAT UNTIL Eot = Y
             OR Eom = Y
        IF Master Key = Transaction Key
            CALL Update
            CALL Get Good Trans
        ELSE
            IF Master Key LESS THAN Transaction Key
                CALL Put New Master
                CALL Get Master
            ELSE /* Master Key GREATER THAN Transaction Key */
                CALL Put Unmatched Transaction
                CALL Get Good Trans
            ENDIF
        ENDIF
    ENDREPEAT

    REPEAT UNTIL Eot = Y
        CALL Put Unmatched Transaction
        CALL Get Good Trans
    ENDREPEAT

    REPEAT UNTIL Eom = Y
        CALL Put New Master
        CALL Get Master
    ENDREPEAT

END Update Master
```

Figure 5-28

Pseudocode for Update Master File module.

true; the other half contains the instruction group executed when the condition is false. Each half is clearly identified near the test condition.

Nassi–Shneiderman charts allow the *nesting* of various constructs, just as flowcharts and pseudocode do. Some examples of nesting appear in Figure 5-31.

The Nassi–Shneiderman chart for the UPDATE MASTER FILE module appears in Figure 5-32.

174

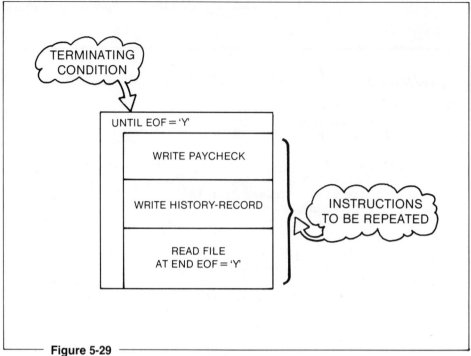

Figure 5-29

Repetition illustrated in Nassi–Shneiderman chart.

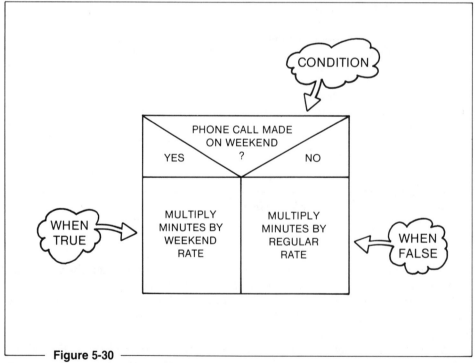

Figure 5-30

Decision illustrated in Nassi-Shneiderman chart.

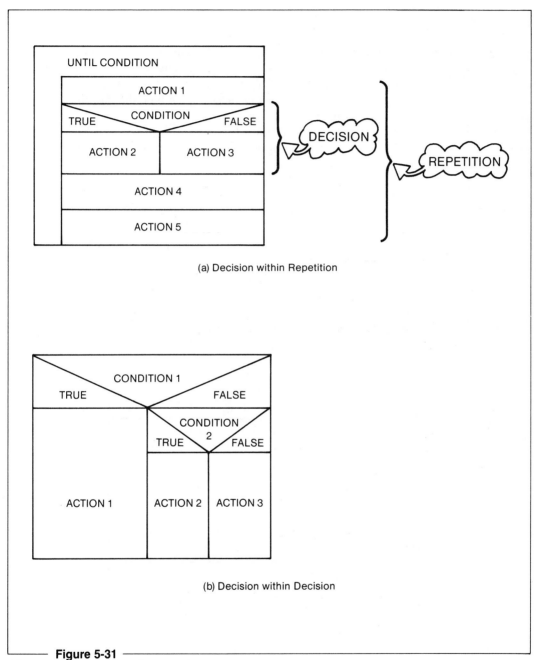

(a) Decision within Repetition

(b) Decision within Decision

Figure 5-31

Nesting in Nassi–Shneiderman charts.

SUMMARY

We specify modules on a structure chart in order clearly and unambiguously to communicate each module's function to the programmer. The module specification is used to guide the programmer who writes the code, and to help future programmers who must read, understand, modify, or replace a module. We can specify modules *nonprocedurally* or *procedurally*.

In this chapter we discussed seven techniques for specifying modules nonprocedurally. All of these specification techniques focus on the module's function

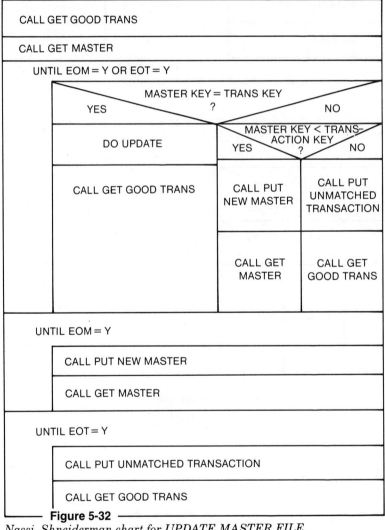

Figure 5-32

Nassi–Shneiderman chart for UPDATE MASTER FILE.

rather than its implementation. The programmer is free to code the module any way he wishes, as long as he adheres to installation coding standards.

Describing input and output parameters is appropriate for modules that perform well-known functions. We simply describe what data the module can expect as input, and the values to which the module must set its output parameters.

Algorithms are useful for specifying modules that perform calculations. This technique assumes that the designer and the programmer are familiar with standard algebraic notation.

Card, tape, and disk layouts are appropriate specification techniques for modules that handle card, tape, and disk files. Those I/O modules are usually found at the bottom of the structure chart, and they require the most details about a file. We can use a general tabular format to describe any type of file, or special preprinted forms for files on specific media. The general format allows greater flexibility, while the special forms are more graphic.

Report or CRT display layouts are two-dimensional grids appropriate for modules that format and write printed reports or CRT displays. We show both constant and variable data on the grid, as well as some editing requirements. Notes should be included to tell the programmer about anything unusual, or to clarify ambiguous points.

Tables are good specification tools for those modules that handle simple conditions. Using a table as a specification technique does not mean that the programmer must code a table within the module.

Decision tables and decision trees are both appropriate specification techniques for describing modules that handle compound conditions. Each technique illustrates all combinations of conditions, and the action or actions that must be taken for each combination (or rule). Although decision tables and decision trees can be used interchangeably, decision tables seem to be more useful when the number of actions is small and several rules result in the same action, and when rules require several actions; decision trees are more useful when the test of one condition depends on the outcome of a previous condition, and when the number of actions is almost as large as the number of rules.

Frequently we use *more than one technique* to specify a single module: the techniques are not mutually exclusive. They can be used in whatever combination best communicates the module's function to the programmer.

Procedural module specifications enable a designer to draft a model of a module's basic logic and review it thoroughly before the actual program instructions are written. Such specifications are very appropriate, even essential, for control modules—ones that appear at the top of a structure chart. They leave little room for interpretation, requiring only that the programmer follow the logic as it is specified. Because of this, procedural specifications can also be used to guide novice programmers.

In writing procedural specifications, a designer accounts for the order in which instructions are executed: they may be executed in a *sequence* (one after another), as the result of a *decision* (one of a set of mutually exclusive instruction groups is executed depending on the result of a test), or they may be *repeated* in a loop until some terminating condition is satisfied. A designer also accounts for *nesting* these control constructs by replacing instruction groups with decisions or loops.

There are three ways to illustrate procedural specifications: flowcharts, pseudocode, and Nassi–Shneiderman charts. A designer is able to indicate sequence, decision, repetition, and nesting using any of these three procedural specification techniques. To illustrate module logic, flowcharts and Nassi–Shneiderman charts use symbols and words, while pseudocode combines keywords and indentation of English-like sentences.

KEY WORDS

Algorithm   A calculation expressed in algebraic notation.

Decision   The point in program logic at which one control path out of a set is selected, based on the result of some test condition.

Decision table   A nonprocedural specification technique that illustrates which activities are to be executed for each unique set of conditions. It uses a two-dimensional table format.

Decision tree   A nonprocedural specification technique that indicates every possible control path through a series of conditions. By tracing a control path we can identify the activities that are to be executed.

Flowchart   A graphic representation of instruction sequences within a program. A procedural specification technique.

Nassi–Shneiderman chart   A graphic representation of instruction sequences within a program. A procedural specification technique.

Nesting   Placing one program control construct within another one.

Nonprocedural specification technique   A method for specifying a module that focuses on the module's function rather than its implementation. See also *procedural specification technique.*

Procedural specification technique   A method for specifying a module that focuses on the sequence of instructions within the module. See also *nonprocedural specification technique.*

Pseudocode   A procedural specification technique for specifying instruction sequences using a vocabulary that strongly resembles, but is not identical to, actual program code.

Repetition   A program control construct in which an instruction group is executed over and over until some terminating condition is met.

Rule   The unique combination of conditions in a decision table or decision tree.

Sequence   A program control construct in which one instruction group is executed after another.

EXERCISES

Refer to the structure chart in Figure 5-E1. Each module has been numbered for reference. The structure chart is taken from the Agency case study, so you can also refer to documents in the appendix.

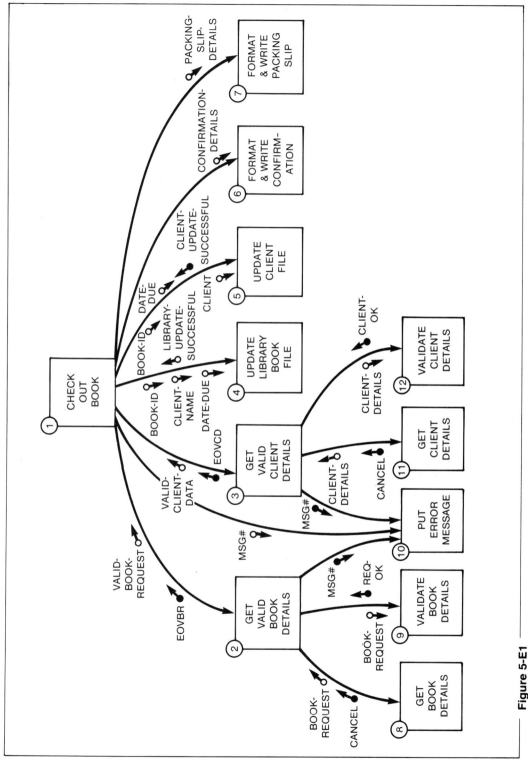

Figure 5-E1
Structure chart for CHECK OUT BOOK.

1. For each module, indicate whether you would specify it nonprocedurally or procedurally (remember that you can combine techniques). Be able to explain your choice.

2. Specify modules 1, 2, and 3 procedurally—
   (a) Draw a flowchart for module 1.
   (b) Draw a Nassi–Shneiderman chart for module 2.
   (c) Write pseudocode for module 3.

3. Specify modules 4 and 5 by describing their input and output parameters, and by including file layouts for the two updated files.

4. Specify module 6 with a CRT display layout.

5. Specify module 7 with a print report layout.

6. Specify modules 8 and 11 by describing their input and output parameters.

7. Specify modules 9 and 12 with lists of validation criteria.

8. Specify module 10 with a list of error-message numbers and error messages.

Refer to the following policy to do the next exercises.

> *Funding Eligibility.* All visually handicapped children (both legally blind and visually impaired) are eligible for State Grant Funds, regardless of their eligibility for any other source of funds. Only legally blind children can receive funding from APH. Children who are visually handicapped and who have a physical, mental, or emotional handicap are entitled to receive funds from the Multihandicapped Account. Any visually handicapped child who is under 3 years 8 months and who is not yet enrolled in kindergarten or a higher grade is eligible for funds from the Preschool Grant Account.

The module that determines funding eligibility is shown in Figure 5-E9.

9. Draw a decision table for the Funding Eligibility policy.

10. Draw a decision tree for the Funding Eligibility policy.

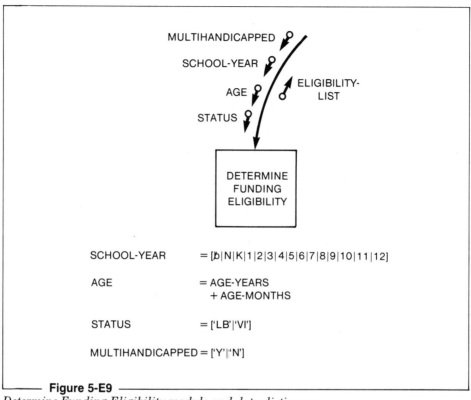

SCHOOL-YEAR = [b̸|N|K|1|2|3|4|5|6|7|8|9|10|11|12]

AGE = AGE-YEARS
      + AGE-MONTHS

STATUS = ['LB'|'VI']

MULTIHANDICAPPED = ['Y'|'N']

Figure 5-E9

Determine Funding Eligibility module and data dictionary.

CHAPTER 6

Reviewing Program Specifications

When you complete this chapter, you will be able to—

- prepare materials for a review
- participate in a review of program specifications

INTRODUCTION

Programs are written according to program specifications. Errors in program specifications become errors in the programs, just as design flaws in a blueprint become flaws in the actual structure. You can imagine how much more expensive and time-consuming it is to alter a building than it is to modify blueprints. That is why a set of blueprints is reviewed before a building is constructed from it. Likewise, it is less expensive and much faster to modify program specifications than it is to modify actual programs. Before program specifications are handed over to a programmer, they should be carefully reviewed in order to identify errors, ambiguities, inconsistencies, and design weaknesses. Reviewing program specifications is an indispensable step in system development, because errors that go undetected now become more difficult and more costly to locate and correct later on.

The objective of a review is to check program specifications for correctness. If a thorough review is to be conducted, a complete set of documents is needed: the program specifications (which include the structure chart and module specifications), the data dictionary, and the process specifications. The structure chart and module specifications are the documents being reviewed; review participants may need the data dictionary and process specifications in order to verify data composition and accuracy of module specifications.

THE REVIEWERS

The expression "Two heads are better than one" never rang as true as in the context of a review. In order to assure some control over the quality of the final product—in this case, the program specifications—someone other than the author of the structure chart and module specifications must give approval. Authors are too close to their products to review them as objectively as someone else can. Therefore, gathering together a group of knowledgeable people in a formal setting to review program specifications according to standard procedures almost ensures that, if there are any errors, they will be found.

The number of review participants can vary considerably, depending on the size and complexity of the program specifications being reviewed. For example, simple report-writing programs require the time and skills of fewer people (two or three, perhaps) than programs that perform lengthy and complex calculations, or interactive programs (five or six reviewers might be needed). Regardless of the number of participants, there are certain roles that must be filled. A *moderator* presides over the meeting. The moderator's job is to keep the participants on task by limiting discussion to the program specifications being reviewed, and to steer the reviewers away from discussions on style or personal preference. A *secretary* keeps a list of all errors the reviewers find. The list is used by the author after the meeting as a reference: the author must correct each and every error that was found. The same list is used in a subsequent review to confirm that all the errors were indeed repaired. The *author's* role during a review is to answer questions, clarify points, and provide explanations when they are needed by the reviewers. The review participants (including the moderator and the secretary) direct their collective energy and expertise toward finding defects. If none are found, the program specifications can be approved; if errors are uncovered, then the author must eliminate all defects and resubmit the specification for another review.

Just as the number of reviewers varies with the size and complexity of the program specifications being reviewed, so does the length of time needed for the review. If too little time is set aside, then reviewers might be rushed and overlook errors in their haste. On the other hand, the longer a review takes, the more costly it is (a one-hour review involving five people who each earn $26,000 per year costs $62.50). Ideally we would like to set aside enough time for a thorough review, and then spend that time productively. The moderator plays an especially important role here by limiting discussion and controlling the meeting.

PREPARING FOR A REVIEW

The moderator also makes the preparations for a review. The moderator must—

notify participants of the date, time, and place of the review;

appoint a participant as secretary;

distribute copies of the program specifications to participants at least two days prior to the review date (to give them enough time to become familiar with—and scrutinize—the documents);

schedule or reserve a meeting room;

obtain additional supplies for the review such as flip chart, marking pens, overhead projector, chalk board, and so on.

CONDUCTING A REVIEW

During a review, participants search for inconsistencies, inaccuracies, errors, and ambiguities in the program specification. They examine the structure chart: module names, parameters passed between modules, organization of the modules, and so forth. They study each module specification, and they compare it to the process specification from which it was derived. They examine the relationship between the module specification and its corresponding module on the structure chart. They scrutinize the data needed by each module, and they

STRUCTURE CHART

Module Names
> Strong verb, specific direct object?
> Accurate statement of what it does for its boss (compare with module specification)?

Organization of Modules
> Does every worker do part of its boss's function?
> Does any boss have more than 7 workers (except transaction centers)?
> Is any module doing too much (refer to module specification)?
> Are higher-level modules shielded from physical data details?

Passed Parameters
> Are all parameters defined in the data dictionary?
> Does each module get all the data it needs (compare module specification to structure chart, noting both the data passed down to the module from its boss, and the data sent up from all its workers)?
> Does each module use all the data it gets, both from above and below?
> Can any flag be eliminated?
> Are there flags missing?
> Is there any flag going down from a boss to a worker? If so, is it absolutely necessary?
> Do names of flags indicate that they are reporting on a situation?
> Do boss and worker modules expect the same settings for flags (refer to module specifications)?

Can any module be made more generally useful?
Will the program be easy to maintain?
Is the design flexible—that is, will it be easy to change?

MODULE SPECIFICATIONS

Is every module specified?
Compare each module specification to the process specification from which it was derived.
> Is it correct?
> Is it complete?
> Is it unambiguous?

Is the module specification technique appropriate for the module function (procedural specification vs nonprocedural specification)?
Does the module have access to all the data it needs, either passed down from a boss or retrieved by a worker (compare to structure chart)?
Is each module capable of producing the output its boss expects it to produce (compare to structure chart)?

Figure 6-1
Review Checklist.

check the data dictionary to verify that the data will be available. They test each module specification with mock data, "playing computer" in an effort to identify any condition the designer neglected to take into account. There is a Review Checklist in Figure 6-1. This checklist is helpful both to authors preparing for a review and to review participants.

As defects are identified in the program specifications, the secretary notes them. Although it is sometimes tempting to do so, review participants should be careful to spend minimal time in actually fixing the errors. It is the author's responsibility to do that after the review. If the error was only an oversight, then pointing it out to the author is sufficient. Otherwise, the review team may make brief suggestions that can be developed later, if and as needed. The moderator is responsible for limiting discussion to error *detection*.

A review is also an inappropriate setting for discussions on style. If there is no error, no ambiguity, no inconsistency, no design weakness, then there is no room for comments such as "Well, I would have done it differently." Sharing ideas about style is a valuable activity, but a review is neither the time nor the place for such discussions. Remember, the goal is to detect all errors in the specifications. All of the energy in a review should be directed toward that goal. Style discussions simply waste valuable time, and should be avoided.

FOLLOW-UP

As a result of the review, the program specifications are approved (no errors were found) or rejected (at least one error was detected). Approved specifications can be given to programmers for implementation. Rejected specifications must be fixed and reviewed again. As noted above, the author—using the list kept by the secretary as a guideline—is responsible for making modifications. The new program specifications, along with the list of errors, are submitted for another review.

AN EXAMPLE

Following are the program specifications for a program that schedules the payment of vendor invoices. It examines each pending invoice and then decides whether it would be more worthwhile to pay the invoice or to delay payment in order to invest the money in a money-market fund. An invoice that is not paid is usually subject to interest, as determined by the vendor. Thus the program compares the amount of money to be earned by investing it against the amount the company would have to pay the vendor in interest. Then it chooses the option that is financially better for the company. This company's policy specifies, however, that all invoices must be paid in 90 days, regardless of the cost.

All vendors with whom the company does business have four interest-rate periods: 0 to 30 days, 31 to 60 days, 61 to 90 days, and over 90 days. Interest rates charged on overdue invoices may vary among periods—and, of course, among vendors. When payments are scheduled, they are scheduled for the last day within the payment period, giving the company use of the money for the longest time. Only payments due within the next 30 days are scheduled for payment. Other invoices are written to a new Pending Invoice file.

Scheduled payments are written on a printed report. An image of a record subsequently used as input to a checkwriting program is also written. The code 416 is assigned to every scheduled payment to identify it as a vendor invoice payment.

Figure 6-2 is the structure chart for the Schedule Invoice Payment program. The data dictionary entries and module specifications appear in Figures 6-3 through 6-5.

You should review the specifications yourself and try to find errors and design problems *before you continue.*

The following points were raised during one review of the SCHEDULE INVOICE PAYMENTS program specifications.

1. ELAPSED-DAYS is an unsigned field. Is it possible to get a negative number for elapsed days between dates? For example, could a vendor issue an invoice that was predated, so that TODAYS-DATE precedes INVOICE-DATE?

2. The module specification for SCHEDULE INVOICE PAYMENTS tests EOPI for 'Y', but the subroutine sets it to '1'.

3. What happens to the return code from GET VENDOR NAME, GET VENDOR ADDRESS, and GET VENDOR TERMS?

4. GET VENDOR NAME, GET VENDOR TERMS, and GET VENDOR ADDRESS could be placed in a subprogram as modules sharing a common storage area in which VENDOR RECORD will be stored.

Points 1, 2, and 4 were recognized as errors or problems, and the specifications were modified as a result (see Figure 6-6). In a discussion of point 3, it was determined that only valid PENDING INVOICES would get to the PENDING INVOICES FILE, so it would be impossible to encounter a missing vendor. Therefore, the return codes do not have to be tested.

With the approved specifications as a guide, a programmer wrote SCHEDULE INVOICE PAYMENTS. Figures 6-7 and 6-8 show the COBOL code for these modules: SCHEDULE INVOICE PAYMENTS, DETERMINE PAYMENT DATE, PUT CHECK RECORD, PUT REPORT, GET VENDOR NAME, GET VENDOR ADDRESS, and GET VENDOR TERMS.

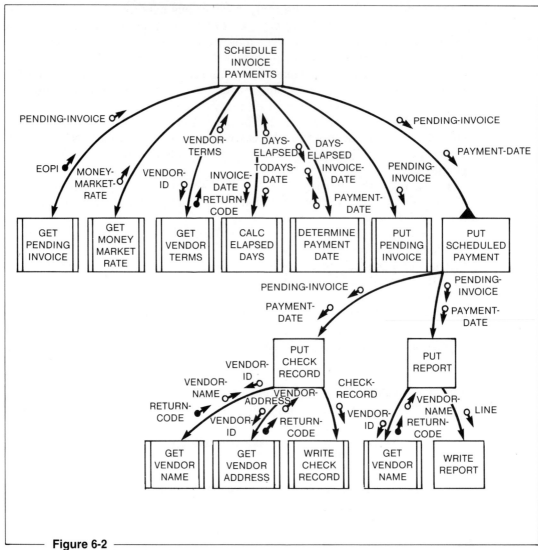

Figure 6-2

Structure chart for the Schedule Invoice Payment program.

PENDING-INVOICE = INVOICE-NUMBER + INVOICE-DATE + VENDOR-ID + AMOUNT-DUE

CHECK-RECORD = PAYMENT-DATE + (INVOICE-NUMBER) + PAYEE-NAME + PAYEE-ADDRESS + CHECK-AMOUNT + CHECK-CODE

VENDOR-RECORD = VENDOR-ID + VENDOR-NAME + VENDOR-ADDRESS + 4 {PERIOD + INTEREST-RATE}

Figure 6-3

Data dictionary for SCHEDULE INVOICE PAYMENT.

CHECK RECORD		Sequential	218 records per block	Disk
POSITION	TYPE	FIELD NAME	COMMENTS	
001 008	N	PAYMENT-DATE	MMDDYYYY	
009 023	A/N	PAYEE-NAME		
024 079	A/N	PAYEE-ADDRESS		
080 084	A/N	INVOICE-NUMBER	Optional	
085 091	N	CHECK-AMOUNT	S99999V99	
092 094	A/N	CHECK-CODE	See list below	

Check Codes List

082	Payroll check
109	Customer refund
416	Vendor invoice
875	T & E

PENDING INVOICE		Sequential	518 records per block	Tape
POSITION	TYPE	FIELD NAME	COMMENTS	
001 005	A/N	INVOICE NUMBER		
006 008	N	VENDOR-ID		
009 016	N	INVOICE DATE	MMDDYYYY	
017 022	N	AMOUNT-DUE		

VENDOR RECORD		Indexed sequential	209 records per block	Disk
POSITION	TYPE	FIELD NAME	COMMENTS	
001 001	A/N	DELETE-CODE	'D' if inactive	
002 004	N	VENDOR-ID	Key field	
005 020	A/N	VENDOR-NAME		
021 076	A/N	VENDOR-ADDRESS		
077 079	N	PERIOD1	Constant '030'	
080 084	N	INTEREST1	S9V9999	
085 087	N	PERIOD2	Constant '060'	
088 092	N	INTEREST2	S9V9999	
093 095	N	PERIOD3	Constant '090'	
096 100	N	INTEREST3	S9V9999	
101 103	N	PERIOD4	Constant '999'	
104 108	N	INTEREST4	S9V9999	

Figure 6-4

Record layouts.

```
BEGIN SCHEDULE INVOICE PAYMENTS

CALL GET MONEY MARKET RATE
CALL GET PENDING INVOICE
DO UNTIL EOPI = "Y"
    CALL CALCULATE ELAPSED DAYS
    IF ELAPSED DAYS LESS THAN OR EQUAL 60
    THEN
        CALL GET VENDOR TERMS
        IF MONEY MARKET RATE GREATER THAN VENDOR INTEREST for
        this period
        THEN
            CALL PUT PENDING INVOICE /*invest the money
            rather than pay */
        ELSE
            CALL DETERMINE PAYMENT DATE /*schedule payment*/
            CALL PUT CHECK RECORD
            CALL PUT REPORT
        ENDIF
    ELSE /*Invoice must be paid - it is over 60 days old*/
        CALL DETERMINE PAYMENT DATE
        CALL PUT CHECK RECORD
        CALL PUT REPORT
    ENDIF
    CALL GET PENDING INVOICE
ENDDO

END SCHEDULE INVOICE PAYMENTS
```

———— **Figure 6-5** ————

Module specifications and report layout.

Module Name: GET VENDOR ADDRESS
Input parameters: VENDOR-ID
Output parameters: Valid vendor:
                        VENDOR-ADDRESS
                        RETURN-CODE = 0
                    Invalid vendor:
                        RETURN-CODE = 1

Module Name: CALCULATE ELAPSED DAYS
Input parameters: EARLY-DATE (mmddyyyy)
                       LATER-DATE (mmddyyyy)
Output parameters: ELAPSED-DAYS (0 to 99999)

Module Name: DETERMINE PAYMENT DATE
Calculate PAYMENT-DATE by adding high value of period range in which
   DAYS-ELAPSED falls to INVOICE-DATE. Ranges are: 0 to 30 days, 31 to 60
   days, 61 to 90 days, 91 to 120 days.

———— **Figure 6-5** *(continued)* ————

Module Name: GET MONEY MARKET RATE
Input parameters: none
Output parameters: MONEY-MARKET-RATE   (9V9999, such that 8½ =
                                          00850 and 100% = 10000)

Module Name: GET VENDOR TERMS
Input parameters: VENDOR-ID
Output parameters: Valid vendor:
                          VENDOR-TERMS
                          RETURN-CODE = 0
                      Invalid vendor:
                          RETURN-CODE = 1

Module Name: GET VENDOR NAME
Input parameters: VENDOR-ID
Output parameters: Valid vendor:
                          VENDOR-NAME
                          RETURN-CODE = 0
                      Invalid vendor:
                          RETURN-CODE = 1

Module Name: GET PENDING INVOICE
Input parameters: none
Output parameters: if not end of file:
                          PENDING-INVOICE
                          EOPI = 0
                      if end of file:
                          EOPI = 1

Figure 6-5 *(continued)*

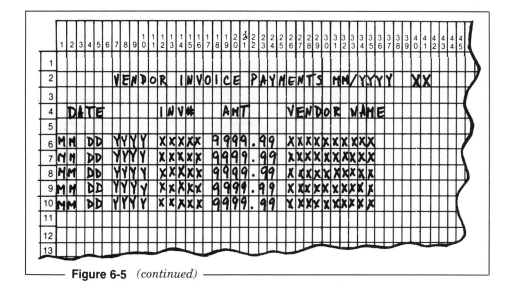

Figure 6-5 *(continued)*

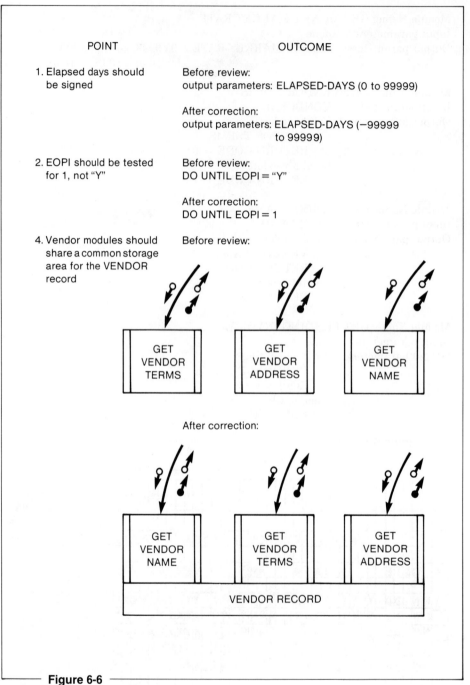

POINT

OUTCOME

1. Elapsed days should be signed

Before review:
output parameters: ELAPSED-DAYS (0 to 99999)

After correction:
output parameters: ELAPSED-DAYS (−99999 to 99999)

2. EOPI should be tested for 1, not "Y"

Before review:
DO UNTIL EOPI = "Y"

After correction:
DO UNTIL EOPI = 1

4. Vendor modules should share a common storage area for the VENDOR record

Before review:

GET VENDOR TERMS

GET VENDOR ADDRESS

GET VENDOR NAME

After correction:

GET VENDOR NAME

GET VENDOR TERMS

GET VENDOR ADDRESS

VENDOR RECORD

Figure 6-6

Modifications to specifications.

```
IDENTIFICATION DIVISION.
PROGRAM-ID.    SCHEDULE.
ENVIRONMENT DIVISION.
CONFIGURATION SECTION.
    C01 IS TOP-OF-PAGE.
INPUT-OUTPUT SECTION.
    SELECT SCHEDULED-PAYMENT-REPORT ASSIGN TO xxxxx
DATA DIVISION.
FILE SECTION.
FD    SCHEDULED-PAYMENT-REPORT.
01    PRINT-RECORD.
      05    CARRIAGE-CONTROL    PIC 9.
      05    PRINT-LINE      PIC X(132).
WORKING-STORAGE SECTION.
01    MONEY-MARKET-RATE   PIC 9V9999.
01    EOPI               PIC 9.
01    PENDING-INVOICE.
      05    INVOICE-NUMBER PIC X(5).
      05    VENDOR-ID      PIC 999.
      05    INVOICE-DATE   PIC 9(8).
      05    AMOUNT-DUE     PIC 9999V99.
01    TODAYS-DATE         PIC 9(8).
01    ELAPSED-DAYS        PIC S9(5).
01    VENDOR-TERMS.
      05    INTEREST-PERIOD OCCURS 4 TIMES.
         10    PERIOD      PIC 999.
         10    INTEREST-RATE   PIC S9V9999.
01    SUBSCRIPT           PIC 9.
01    JULIAN-DATE.
      05    JULIAN-YEAR    PIC 99.
      05    JULIAN-DAY     PIC 999.
01    SCHEDULED-PAYMENT-DATE   PIC 9(8).
01    VENDOR-NAME         PIC X(16).
01    RETURN-CODE-VT      PIC 9.
01    RETURN-CODE-VN      PIC 9.
01    RETURN-CODE-VA      PIC 9.
01    FIRST-TIME-SWITCH   PIC 9      VALUE 1.
01    LINE-COUNT          PIC 999.
01    CHECK-RECORD.
      05    PAYMENT-DATE   PIC 9(8).
      05    PAYEE-NAME     PIC X(15).
      05    PAYEE-ADDRESS  PIC X(57).
      05    INVOICE-NUMBER PIC X(5).
      05    CHECK-AMOUNT   PIC S99999V99.
      05    CHECK-CODE     PIC XXX.
```

Figure 6-7

COBOL code for SCHEDULE INVOICE PAYMENTS. Code is continued on following pages.

```
01    SCHEDULED-PAYMENT-LINE.
      05   FILLER          PIC 9      VALUE 1.
      05   PR-PAYMENT-DATE    PIC 99B99B9999.
      05   FILLER          PIC X      VALUE SPACE.
      05   PR-INVOICE-NUMBER   PIC X(5).
      05   FILLER          PIC X      VALUE SPACE.
      05   PR-AMOUNT       PIC ZZZ9.99.
      05   FILLER          PIC X      VALUE SPACE.
      05   PR-VENDOR-NAME PIC X(10).
      05   FILLER          PIC X(46).
01    PAGE-COUNT          PIC 99        VALUE 0.
01    PAGE-HEADER.
      05   FILLER          PIC 9         VALUE 2.
      05   FILLER          PIC X(6)  VALUE SPACE.
      05   FILLER          PIC X(24)
                VALUE 'VENDOR INVOICE PAYMENTS '.
      05   REPORT-DATE     PIC X(7).
      05   FILLER          PIC XX        VALUE SPACE.
      05   PAGE-NUMBER     PIC 99.
01    COLUMN-HEADERS.
      05   FILLER          PIC 9         VALUE 2.
      05   FILLER          PIC X(11)     VALUE 'DATE'.
      05   FILLER          PIC X(7)      VALUE 'INV#'.
      05   FILLER          PIC X(7)      VALUE 'AMT'.
      05   FILLER          PIC X(11)     VALUE 'VENDOR NAME''.

PROCEDURE DIVISION.
   CALL 'MMRATE' USING MONEY-MARKET-RATE.
   CALL 'GETPI' USING PENDING-INVOICE, EOPI.
   MOVE CURRENT-DATE TO TODAYS-DATE.
   PERFORM SCHEDULE-PAYMENTS UNTIL EOPI = 1.
   STOP RUN.

SCHEDULE-PAYMENTS.
   CALL 'ELAPDAY' USING INVOICE-DATE, TODAYS-DATE,
   ELAPSED-DAYS.

   IF ELAPSED-DAYS IS LESS THAN OR EQUAL TO 60
   THEN
       PERFORM INVEST-OR-PAY-INVOICE
   ELSE
       PERFORM DETERMINE-PAYMENT-DATE
       PERFORM PUT-CHECK-RECORD
       PERFORM PUT-REPORT.

   CALL 'GETPI' USING PENDING-INVOICE, EOPI.
```

Figure 6-7 *(continued)*

```
INVEST-OR-PAY-INVOICE.
    CALL 'GETVT' USING VENDOR-ID, VENDOR-TERMS, RETURN-CODE-VT.

    IF ELAPSED-DAYS IS LESS THAN OR EQUAL TO 30
    THEN
        MOVE 1 TO SUBSCRIPT
    ELSE
        MOVE 2 TO SUBSCRIPT.

    IF MONEY-MARKET-RATE IS GREATER THAN INTEREST-RATE
      (SUBSCRIPT)
    THEN
        CALL 'PUTPI' USING PENDING-INVOICE
    ELSE
        PERFORM DETERMINE-PAYMENT-DATE
        PERFORM PUT-CHECK-RECORD
        PERFORM PUT-REPORT.

DETERMINE-PAYMENT-DATE.
    CALL 'GREGJUL' USING INVOICE-DATE, JULIAN-DATE.

    IF ELAPSED-DAYS IS LESS THAN OR EQUAL TO 30
    THEN
        ADD 30 TO JULIAN-DAY
    ELSE
        ADD 60 TO JULIAN-DAY.

    IF JULIAN-DAY IS GREATER THAN 365
    THEN
        SUBTRACT 365 FROM JULIAN-DAY
        ADD 1 TO JULIAN-YEAR.

    CALL 'JULGREG' USING JULIAN-DATE, SCHEDULED-PAYMENT-DATE.

PUT-CHECK-RECORD.
    CALL 'GETVN' USING VENDOR-ID, VENDOR-NAME, RETURN-CODE-VN.
    CALL 'GETVA' USING VENDOR-ID, VENDOR-ADDRESS, RETURN-CODE-VA.

    MOVE INVOICE-NUMBER OF PENDING-INVOICE TO
      INVOICE-NUMBER OF CHECK-RECORD.
    MOVE VENDOR-NAME TO PAYEE-NAME.
    MOVE VENDOR-ADDRESS TO PAYEE-ADDRESS.
    MOVE AMOUNT-DUE TO CHECK-AMOUNT.
    MOVE SCHEDULED-PAYMENT-DATE TO PAYMENT-DATE.
    MOVE '416' TO CHECK-CODE.
```

Figure 6-7 *(continued)*

```
        CALL 'WRTCR' USING CHECK-RECORD.

PUT-REPORT.
    IF FIRST-TIME-SWITCH = 1
    THEN
        OPEN OUTPUT SCHEDULED-PAYMENT-REPORT
        MOVE 0 TO FIRST-TIME-SWITCH
        MOVE 0 TO LINE-COUNT
        PERFORM PRINT-PAGE-HEADERS.

    IF LINE-COUNT IS GREATER THAN OR EQUAL TO 55
    THEN
        PERFORM PRINT-PAGE-HEADERS.

    CALL 'GETVN' USING VENDOR-ID, VENDOR-NAME, RETURN-CODE-VN.

    MOVE SCHEDULED-PAYMENT-DATE TO PR-PAYMENT-DATE.
    MOVE INVOICE-NUMBER TO PR-INVOICE-NUMBER.
    MOVE AMOUNT-DUE TO PR-AMOUNT.
    MOVE VENDOR-NAME TO PR-VENDOR-NAME.

    MOVE SCHEDULED-PAYMENT-LINE TO PRINT-RECORD.
    ADD CARRIAGE-CONTROL TO LINE-COUNT.
    PERFORM WRITE-LINE.

PRINT-PAGE-HEADERS.
    ADD 1 TO PAGE-COUNT.
    MOVE PAGE-COUNT TO PAGE-NUMBER.
    MOVE SPACES TO PRINT-LINE.
    WRITE PRINT-RECORD AFTER ADVANCING TOP-OF-PAGE.
    MOVE PAGE-HEADER TO PRINT-RECORD.
    PERFORM WRITE-LINE.
    MOVE COLUMN-HEADERS TO PRINT-RECORD.
    PERFORM WRITE-LINE.
    MOVE SPACES TO PRINT-LINE.
    MOVE 1 TO CARRIAGE-CONTROL.
    PERFORM WRITE-LINE.
    MOVE 4 TO LINE-COUNT.

WRITE-LINE.
    WRITE PRINT-RECORD AFTER ADVANCING
        CARRIAGE-CONTROL LINES.
```

Figure 6-7 *(continued)*

```
IDENTIFICATION DIVISION.
PROGRAM-ID. VENDORIO.
ENVIRONMENT DIVISION.
INPUT-OUTPUT SECTION.
FILE-CONTROL.
    SELECT VENDOR ASSIGN TO xxxxx
            ACCESS IS RANDOM
            ACTUAL KEY IS VENDOR-ID
            NOMINAL KEY IS PASSED-VENDOR-ID.
DATA DIVISION.
FILE SECTION.
FD  VENDOR
    BLOCK CONTAINS 209 RECORDS
    LABEL RECORDS ARE STANDARD.
01  VENDOR-RECORD.
    05    DELETE-CODE    PIC X.
    05    VENDOR-ID      PIC 999.
    05    VENDOR-NAME    PIC X(16).
    05    VENDOR-ADDRESS PIC X(57).
    05    VENDOR-TERMS.
        10    PERIOD-RATE OCCURS 4 TIMES.
            15    PERIOD    PIC 999.
            15    INTEREST  PIC S9V9999.

WORKING-STORAGE SECTION.
01  FIRST-TIME-SWITCH  PIC 9      VALUE 0.
01  WORK-VENDOR-RECORD.
    05    WORK-DELETE-CODE    PIC X.
    05    WORK-VENDOR-ID PIC 999.
    05    WORK-VENDOR-NAME    PIC X(16).
    05    WORK-VENDOR-ADDRESS PIC X(57).
    05    WORK-VENDOR-TERMS.
        10    WORK-PERIOD-RATE OCCURS 4 TIMES.
            15    WORK-PERIOD    PIC 999.
            15    WORK-INTEREST  PIC S9V9999.
01  READ-SUCCESSFUL    PIC X.

LINKAGE SECTION.
01    PASSED-VENDOR-ID    PIC 999.
01    RETURN-VENDOR-TERMS PIC X(32).
01    RETURN-VENDOR-NAME  PIC X(16).
01    RETURN-VENDOR-ADDRESS    PIC X(57).
01    NO-SUCH-VENDOR-FLAG PIC 9.
```

Figure 6-8

COBOL code for Get Vendor Terms, Get Vendor Name, and Get Vendor Address modules.

```
PROCEDURE DIVISION.
READ-VENDOR-RECORD.
    IF FIRST-TIME-SWITCH = 1
    THEN
        OPEN INPUT VENDOR
        MOVE 0 TO FIRST-TIME-SWITCH.

    MOVE 'Y' TO READ-SUCCESSFUL.
    READ VENDOR INTO VENDOR-WORK-AREA
        INVALID KEY MOVE 'N' TO READ-SUCCESSFUL.

ENTRY 'GETVN' USING PASSED-VENDOR-ID
              RETURN-VENDOR-NAME
              NO-SUCH-VENDOR-FLAG.
    IF PASSED-VENDOR-ID NOT EQUAL WORK-VENDOR-ID
    THEN
        PERFORM READ-VENDOR-RECORD
        IF READ-SUCCESSFUL = 'N'
        THEN
            MOVE 1 TO NO-SUCH-VENDOR-FLAG
        ELSE
            NEXT SENTENCE
    ELSE
        MOVE WORK-VENDOR-NAME TO RETURN-VENDOR-NAME
        MOVE 0 TO NO-SUCH-VENDOR-FLAG.
    GOBACK.

ENTRY 'GETVA' USING PASSED-VENDOR-ID
                RETURN-VENDOR-ADDRESS
                NO-SUCH-VENDOR-FLAG.
    IF PASSED-VENDOR-ID NOT EQUAL TO WORK-VENDOR-ID
    THEN
        PERFORM READ-VENDOR-RECORD
        IF READ-SUCCESSFUL = 'N'
        THEN
            MOVE 1 TO NO-SUCH-VENDOR-FLAG
        ELSE
            NEXT SENTENCE
    ELSE
        MOVE WORK-VENDOR-ADDRESS TO
          RETURN-VENDOR-ADDRESS
        MOVE 0 TO NO-SUCH-VENDOR-FLAG.
    GOBACK.
```

Figure 6-8 *(continued)*

```
ENTRY 'GETVT' USING PASSED-VENDOR-ID
                    RETURN-VENDOR-TERMS
                    NO-SUCH-VENDOR-FLAG.
    IF PASSED-VENDOR-ID NOT EQUAL TO WORK-VENDOR-ID
    THEN
        PERFORM READ-VENDOR-RECORD
        IF READ-SUCCESSFUL = 'N'
        THEN
            MOVE 1 TO NO-SUCH-VENDOR-FLAG
        ELSE
            NEXT SENTENCE
    ELSE
        MOVE WORK-VENDOR-TERMS TO
          RETURN-VENDOR-TERMS
        MOVE 0 TO NO-SUCH-VENDOR-FLAG.
    GOBACK.
```

Figure 6-8 *(continued)*

KEY WORDS

Author The individual who developed the document being reviewed.
Moderator The review participant who makes preparations for the review, and then conducts it.
Review A critical assessment of docu-ments by a group of knowledgeable people. The objective of a review is to check doc-uments for correctness.
Secretary The review participant who takes notes during a review.

EXERCISES

1. Using one of the structure charts you drew for the Exercises at the end of Chapter 4, prepare for and conduct a review. Pick three or four classmates as participants.

2. In teams of four to five students, draft program specifications for Diagram 4 in the appendix, Receive Book in Resource Center. When your team is satisfied with its product, exchange documents with another team and review each other's specifications.

3. Refine the structure chart in Figure 6-E3. Using the data dictionary and process specifications in the appendix, write module specifications. Then present your program specification documents to a group of classmates for review.

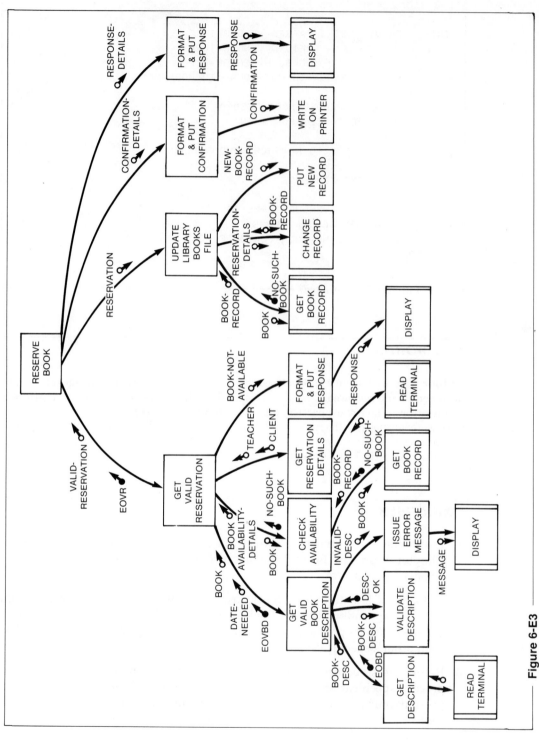

Figure 6-E3

First-cut structure chart for RESERVE BOOK.

PART 2

Other System Components

CHAPTER 7

Organizing and Storing Data

When you complete this chapter, you will be able to—

- describe application criteria that influence data-organization and storage decisions

- distinguish between file systems and database-management systems

- compare data-storage devices in terms of capacity, speed of data retrieval, and limitations

INTRODUCTION

What are data? Data are recorded facts and figures, the stuff upon which computer systems are built. In fact, it is because computers can process data so quickly and can store so many data for easy access that we have computer systems at all. In this chapter, we examine *stored data*. We consider storage devices, file organization, and retrieval. Then we examine database-management systems as an alternative for storing and retrieving data.

STORING DATA

We record facts and figures about things so that we can refer to them in the future. For example, at the Agency for the Blind, we might want to be able to look up a client's address, telephone number, and date of birth. We might need to find a teacher's home telephone number. We might want to know if a brailled copy of a particular book is available for a student. Therefore we record certain facts about each and every client: name, address, phone number, eye condition, date of birth, reading modes. A client record might look like the one in Figure 7-1. We also record certain facts about each teacher: name, address, home telephone number, and state certification number (see Figure 7-2). And of course we record data about books, expenditures, and funding sources.

In addition to retrieving facts and figures about particular things, we sometimes need to look up *relationships* between things. For example, a relationship between a client and a teacher is established when the client is assigned to the teacher's caseload. We might want to find out the name of a particular client's teacher. Or we might want to know the names of all the clients assigned to one teacher. Thus, when we store data for future use, we must have a means of keeping track of the relationships among different objects.

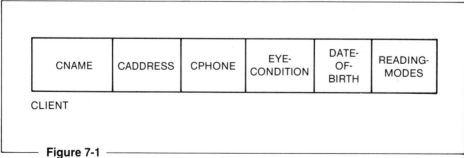

Figure 7-1

Client record.

Figure 7-2
Teacher record.

How do we illustrate the relationship between client and teachers? Figure 7-3 shows one way: draw a line leading from one record to the other. That line means that we'll have to figure out a way to store the data about clients and teachers so that if we know the teacher we can find out who his clients are, and if we know the client we can find out who his teacher is.

Let's expand this example by adding another object from the Agency, the books in the Resource Center. Facts concerning each book include the title, copyright date, author, and medium (see Figure 7-4). Now we can add the book

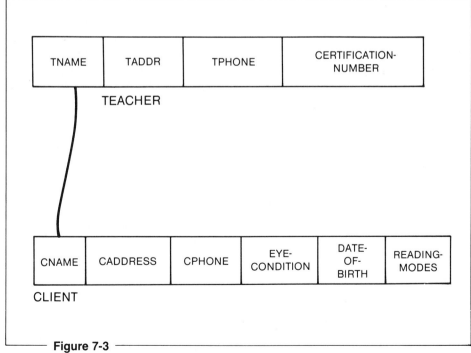

Figure 7-3
Client-teacher relationship.

record to the combination of client and teacher records from Figure 7-3 in order to illustrate the relationship established between clients and books when a book is checked out of the Resource Center (see Figure 7-5).

Now we will move ahead and illustrate the entire set of related records for the Agency. In Figure 7-6, we add the following objects:

funding-source records, which contain funding source name, starting balance, and current balance (funding sources are related to clients because each funding source can be used by many clients; funding sources are related to expenditures because they provide funds for expenditures);

expenditures, which have descriptions and amounts (each expenditure is related to the client for whom it was made, and to a funding source that provided the money to pay for it).

Figure 7-6 is getting pretty crowded. If we needed to add more types of objects (we don't, but *if* we did), then the picture might become difficult to read. So let's consider using some shorthand that will illustrate only the relationships among objects, without showing us any of the details about the data fields in each record. Figure 7-7 shows one way.

Notice that the lines representing relationships among objects have arrowheads. The arrow between teacher and client is one-headed at one end and double-headed at the other. This means that a teacher can be associated with many clients, but that a client can be associated with only one teacher. Similarly, a client can sign many books out of the Resource Center, but a book can be issued to only one client. However, a funding source can be used for many clients, and each client may be eligible for funds from many funding sources.

Figure 7-7 summarizes the relationships that may exist between objects, but it does not reveal *how* to represent them physically when we create our data

TITLE	AUTHOR	COPYRIGHT	MEDIUM

BOOK

Figure 7-4
Book record.

files, or how to wend our way through the stored data to answer such questions as the following:

Who is the teacher of the client who checked out a particular book?

How many clients of a particular teacher are eligible for funding from a particular funding source?

How much money has been allocated from a particular funding source for expenditures for students from a particular town?

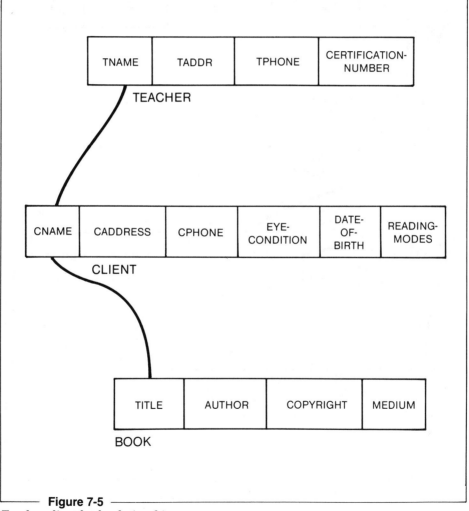

Figure 7-5

Teacher-client-book relationships.

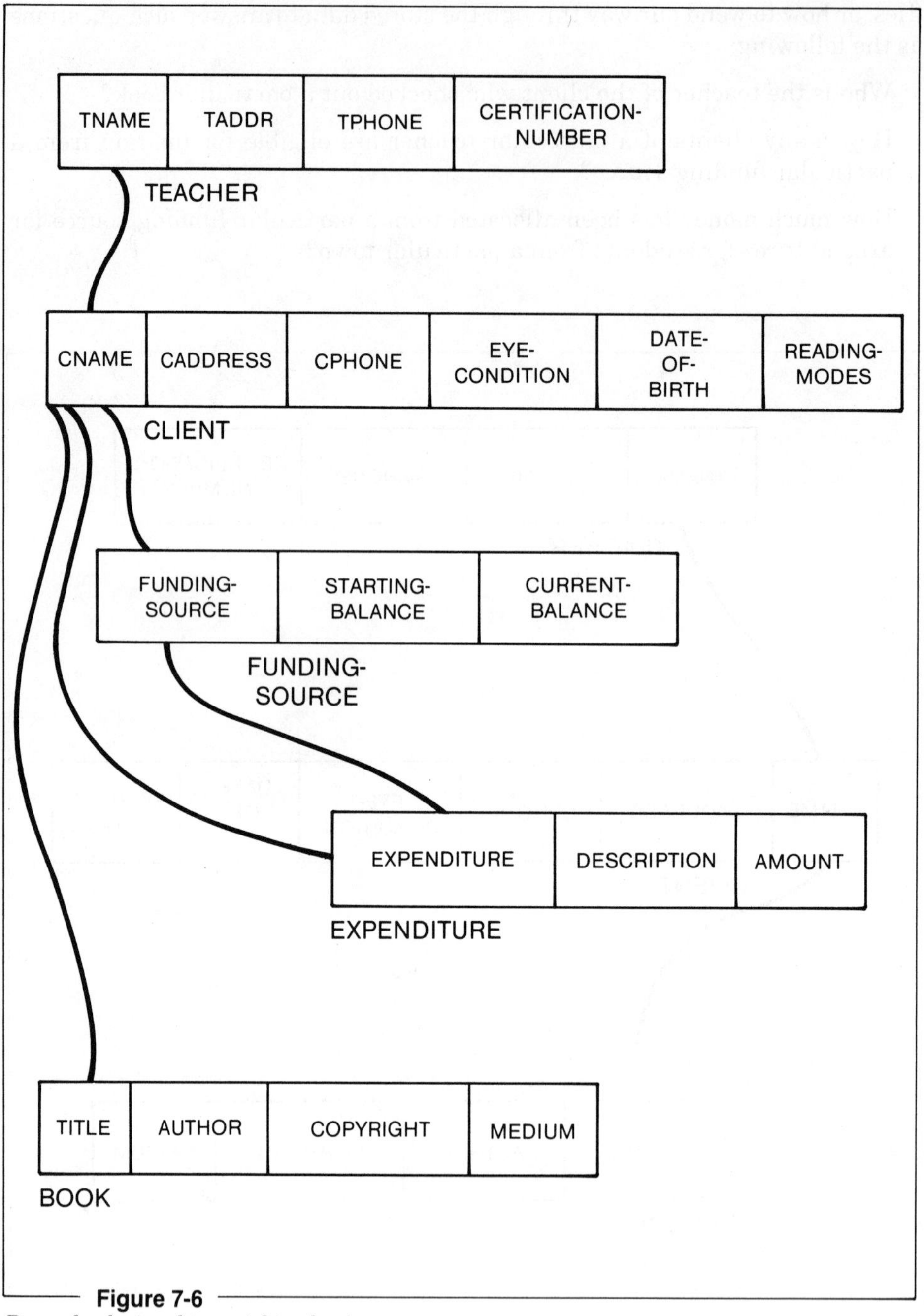

Figure 7-6

Record relationships within the Agency.

In order to answer such questions, we must—

1. organize the data in a particular physical format, called *file organization,* on the storage device;

2. select *access methods* that enable us to retrieve and update the stored data;

3. decide whether we will *write our own programs* to store the data and then retrieve and update it, or use *commercial software* to do the data accessing for us (we can choose either a file-management system or a database-management system).

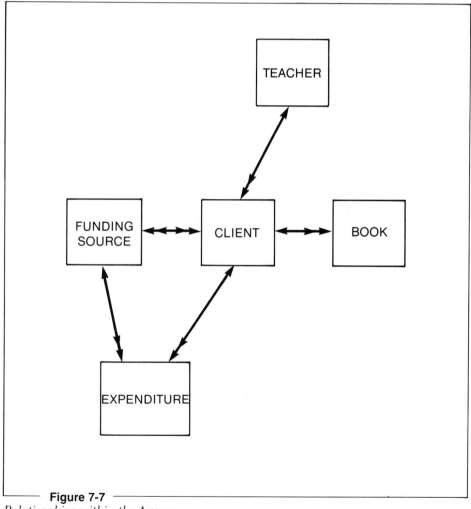

Figure 7-7

Relationships within the Agency.

Figure 7-8 summarizes the options available to us for storing data. In the rest of this section, we examine each option and then establish some guidelines, or rules of thumb, that will help you decide which options are appropriate for a specific system. But before we address any of those options, we must know some more facts about the data we need to store. The facts include answers to the following questions.

How quickly do we need to access specific records?

How many records are we going to store?

How often do we add records to and delete records from the files, and how much of the file do we process at a time?

All of these answers are totally dependent upon the application. And all the answers will have some bearing on the decisions we make regarding devices, access methods, file organization, and databases. So let us first examine the application-dependent variables.

APPLICATION-DEPENDENT VARIABLES

AMOUNT OF DATA Some companies maintain very few records in their files, perhaps only a few thousand of them. A small business serving a few hundred customers or a school with only 5000 students is a relatively small operation (financially it may be very big, but we are considering only data records right now). Conversely, a manufacturing firm employing 8000 workers, maintaining an inventory of 250,000 unique parts, dealing with 10,000 customers, operating 50 plants, purchasing raw materials from 2500 vendors, and communicating with 1500 stockholders represents a vastly larger data-storage

Storage devices	Magnetic tape Magnetic disk Floppy disk, or diskette
File organization	Sequential Indexed sequential Direct
Access methods	Sequential Random, or direct
Programming	Build in-house Buy commercial file-management or database-management system

Figure 7-8
Data storage options.

situation. The sheer number of records to be stored affects both hardware and software decisions (and, of course, decisions regarding people and procedures as well).

SPEED OF RECORD RETRIEVAL   In some applications, the speed with which particular records must be accessed is critical. In certain life-threatening situations (hospital patient monitoring, air traffic control, police work) there is no time to casually peruse a data file looking for a record: the data sought must be retrieved quickly, or grave consequences may result. Other applications are not life-threatening, but still require rapid access to particular records. Bank tellers have almost instant access to customer-account data in order to serve the customer better (and a satisfied customer leaves his money in the bank); telephone information operators have rapid access to directory information; library catalog systems save users countless hours of digging through the card catalog. Both data-storage hardware and software decisions must take the application's retrieval requirements into account.

ANTICIPATED REFERENCE TO STORED DATA   We mentioned above that data are stored so we can reference them again and again in the future. Referencing the stored data—whether by reading, adding new records, deleting obsolete records, or updating—is another variable in determining the hardware and software most appropriate for the application.

Because the collection of stored data represents the state of some enterprise at a particular moment, it is obsolete almost as soon as it is recorded. Consider, as an analogy, a photograph taken of yourself. If the photo is taken today, it may accurately represent your physical appearance. That same photo will not be so accurate next year. In ten years, it will be even more out of phase with reality: your hair may begin to gray, wrinkles may develop around your eyes, or you may have orthodontic work done. Thirty years from now, it might be difficult to tell that it is a photograph of you. The problem is that pictures are permanent, although their subjects change (except for Dorian Gray, of course). If you want an accurate picture of yourself, you must frequently take new ones to keep up with the changes.

The same is true of stored data. The data collection is accurate only for a very short time. If we want it to represent reality, then we have to keep updating it to reflect changes that have already taken place in the real world.

The subjects we will consider here are *activity, volatility, redundancy,* and *data interdependence.*

Activity refers to the percentage of the stored data records that are accessed during a single run. (A run is a unit of computer work, usually the same as a job.) In the Agency-for-the-Blind case study, the number of client records that are updated with expenditures during any one day is relatively small. A telephone information operator references only a fraction of the records available. Conversely, a system that prints out mailing labels from potential customer files reads a large percentage of the stored data each time it is run (probably all of it). And at the end of a semester, a large part of the student-grades file is updated with new grades.

Volatility refers to the percentage of new records added to an existing data file, and to the percentage of records deleted from the file. A volatile file is subjected to a high number of additions and deletions, while a static file is not. Stock-market transaction files are constantly changing as shares are made available (and added to the file) and are purchased (and deleted from the file). Frequent additions and deletions make a file highly volatile. The Agency-for-the-Blind client file, on the other hand, is relatively static. Clients are only occasionally added and deleted.

Redundancy refers to the number of places in which the same piece of data is stored within an enterprise's stored data. Data redundancy can cause problems for an enterprise—in terms of the outsider's perception of the enterprise, and in terms of the ease (or difficulty) with which the enterprise can process and update its data. Consider a bank with three operating departments: savings, checking, and loans. Let's say that each department maintains its own account records. Many customers do business with all three departments, and thus their names and addresses appear in records in all three departments. This is called redundancy. What are some problems with redundancy?

The task of updating a redundant piece of data becomes difficult; because it is difficult, it is frequently done incorrectly or incompletely, resulting in records that are not accurate reflections of the real world. Suppose a customer of the above-mentioned bank moves, informing the bank of the change in address, but somehow only two of the three departments manage to make the change. One department continues to send its notices to the customer's old address. From the customer's point of view, the bank has some real problems: if it cannot even keep an address straight, what is the bank doing with the money?

Another way to think of redundancy is to consider how many users of the stored data need access to a particular piece of data. If "popular" data items can be identified and stored in such a way that all users who want them can access them, then many of the problems with data redundancy can be eliminated.

Data interdependence refers to the constraints placed upon one piece of data by another. For example, in the Agency for the Blind, each client is associated with a teacher. However, the teacher must be on the list of Agency teachers. Expenditures are made on behalf of clients, and are paid for with funds in a funding-source account. The constraints are that the funding source used for an expenditure for a client (1) must be one for which the client is eligible, and (2) must have a balance large enough to cover the expenditure. When a teacher leaves the Agency, the teacher's record will be deleted, but not until all clients have been assigned to another teacher.

When processing one piece of data requires access to another piece of data, they are dependent upon each other. The nature and complexity of the interdependence has an influence on the approach the designer takes when selecting data-storage software. The more complex the interdependence, the more complex the programs necessary to store, maintain, and retrieve the data.

Let us now look at the physical characteristics of storage devices.

PHYSICAL DATA-STORAGE DEVICES

In this section, we examine three popular data-storage devices: magnetic tape, magnetic disk, and floppy disk (or diskette). First we discuss some general characteristics of the three storage media and compare them to one another. We see how to use blocking to make the best use of the storage device and to influence file-processing time.

GENERAL CHARACTERISTICS

Magnetic tape allows several hundred characters to be stored in each inch of recording surface, commonly 800 bytes per inch (bpi) or 1600 bpi. A single reel of tape is usually 2400 feet long, so a reel can hold from 25 million to 50 million characters. Magnetic tape is extremely portable and relatively inexpensive compared to magnetic disk. It must be handled carefully but is pretty rugged nonetheless. It is reusable, of course. The mechanical speed with which the tape passes the read/write heads is slow compared to that possible with diskette and disk.

Magnetic disk is very dense—that is, the number of characters that can be stored on a disk is high. That number depends on the number of recording surfaces as well as the density of the magnetic material on the surfaces. Although there is no standard number, disks can usually store anywhere from 10 million to several hundred million bytes of data.

Removable disks are very heavy, large, and difficult to transport. They must be handled carefully to avoid problems that dust causes when the disk is being accessed. Disks can also be *fixed* (that is, permanently installed) thereby eliminating handling problems. Disks are mounted on disk drives that spin them around very rapidly (from 3000 to 5000 rpm), causing the data to pass under (or over) read/write heads. Each recording surface has its own set of read/write heads. Thus, it does not take long to locate and read data records on a disk file. In fact, disks are called direct-access storage devices (DASDs) because specific records can be located very quickly.

Disks are expensive when compared to tape and diskette. Prices range from several hundred dollars to several thousand.

Floppy disks, or *diskettes,* resemble 45-rpm records permanently sealed in envelopes. Like magnetic tape they are inexpensive (from $2 to $5 per diskette), portable, and easy to handle. Like magnetic disk, diskettes are direct-access storage devices. Storage capacity varies depending on the manufacturer, but a diskette generally holds from 200,000 to 500,000 characters, far less than a reel of tape or a magnetic disk.

Figure 7-9 compares the three storage media.

BLOCKING RECORDS

If a commercial file-management or database-management system is used, then record blocking is handled automatically by the system. If you write your own file-handling software it is important to understand the concept of blocking records.

Data stored on magnetic tape, magnetic disk, or diskette can be *blocked* (see Figure 7-10). A block is the group of data that is transferred from the external device—in this case, the tape drive or disk drive—to the central processor at one time. The most time-consuming part of reading and writing data on external storage devices is the mechanical part—actually activating the drive, physically moving the tape or disk past the read/write heads, and finding the data. Once the data are in the computer, moving them around is electronic rather than mechanical, and thus very rapid. Blocking data records allows us to transfer a group of records to the central processor all at once, rather than one at a time.

Think of blocking as the same technique you use to get your laundry upstairs from the basement laundry room. It takes time to run up and down the stairs. If you had lots of energy to burn off, you might make a separate trip for each towel: bring it upstairs, fold it, put it away, then go down for the next one, and

	Tape	Disk	Diskette
Density	Low	Very high	High
Speed	Slow	Extremely fast	Usually faster than tape, slower than disk
Cost	Cheap	Expensive	Cheap

Figure 7-9

Comparison of storage media.

continue until you finished the laundry and worked off thirty pounds. Or you could take a basket with you and pick up several items at once, bring them all upstairs, fold and put away that batch, and return for another load. You would save yourself quite a bit of time and energy. A block of records is simply the basketful that goes from the external device to the processor at once.

If we decide to block records on the file, then we specify the number of records in a block when we create the file. In order to determine the optimum blocksize, we consider—

the capacity of the track (for disk files);

the amount of space available in main computer memory for input and output buffers;

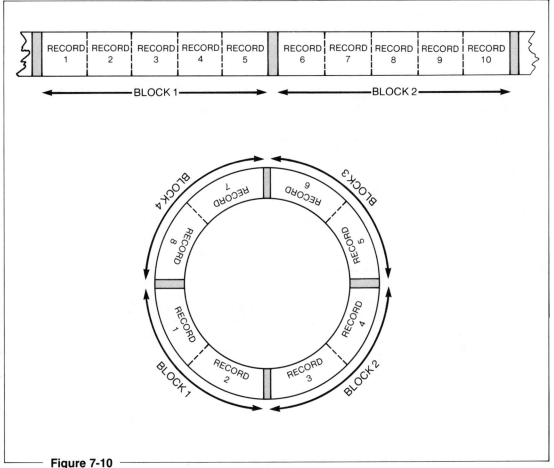

Figure 7-10

Blocked records on tape and disk.

the amount of time it takes to process each record, in conjunction with the number of programs that will be waiting to execute at the same time as the program processing this file.

SIZE OF THE TRACK   A *track* is a concentric circle on the recording surface of a disk that holds a fixed number of characters. The number varies from one disk model to another. Ideally, we would like a single block to occupy one whole track. It would appear simple to determine the number of records that fit on a track; one might simply divide the number of characters that can be stored on a track by the number of characters in a record. The quotient would be the *blocking factor* (number of records per block). Unfortunately, it is not quite so simple. Some characters on the track are used by the operating system or by the file-accessing software. Therefore, they are not available for storing data records.

Because the system software uses some of the storage locations on the disk, and because the number of characters used varies from one file to another, most disk manufacturers publish tables that the designer can use. The tables take into account the characters that the system reserves for its own use. The designer simply turns to the table for a particular disk model, looks for the entry for the desired record size, and reads the optimum blocksize from the table.

In addition, there is a difference between the number of characters reserved for system use with sequential files and with files organized for access by key field. File organization will be discussed shortly. At this point, let us simply point out that files organized for access by key field require more overhead— that is, the system software uses more of the available characters on the track for key-accessible files than for sequential files. That is also figured into the blocksize tables. See Figure 7-11.

Some computer systems used *fixed-size blocks*—that is, no matter what the size of the record, every block always contains the same number of characters. The entire block is not used, unless some number of records happens to match the fixed blocksize. Unused storage locations in the block are bypassed by the system when the record is processed.

BUFFER SIZE   Buffers are holding areas for data waiting to be processed and for data waiting to be written out to a device. The amount of space available for buffers varies from one computer system to another. We cannot put more records in a block than would fit in the buffer available for it.

PROCESSING TIME AND COMPETING PROGRAMS   Another variable to consider when selecting the optimum blocksize for a file is the number of application programs that will be vying for the computer resources at the same time as the programs that process this file.

In many operating systems, programs relinquish control of the computer resources when input/output (I/O) interrupts occur. I/O interrupts occur only when an entire block of input data records has been processed and the program needs a new block, or when an entire block of output records has been written and the block is to be recorded on the external device. Until this happens, the program remains in control of the computer resources, making all the other programs wait their turn.

The amount of processing time spent on each record depends completely upon the application. Some applications process a record with only a few thousand instructions. Complex algorithms, however, can take minutes at a time. If we were to store a large number of records in a block and take some minutes to process each record, then the other programs would have to wait too long for their turns at the computer resources. Of course, the number of other programs waiting in line also depends on the installation.

For most business applications, the amount of processing time per record is small, while the number of records is very high. Therefore, the optimum block-size is less a function of processing time required than a function of track size and buffer limitations. Conversely, in scientific applications, the number of input and output records is relatively small, while the amount of time spent on each record is long. Thus the best blocksize might be one record per block, or at most only a few records per block. When we record only one record in a block, we say the file is *unblocked*.

This many bytes per block—				—will allow you to put this many blocks per		
Without Keys		With Keys				
Min	Max	Min	Max	Track	Cylinder	Pack
6448	13030	6392	12974	1	19	7676
4254	6447	4198	6391	2	38	15352
3157	4253	3101	4197	3	57	23028
.						
.						
.						
464	491	408	435	21	399	161196
438	463	382	407	22	418	168872
414	437	358	381	23	437	176548
.						
.						
.						
36	38			76	1444	583376
34	35			77	1463	591052
.						
.						
.						

Figure 7-11

Part of a disk storage capacity table.

DISKETTE   Data stored on diskette are recorded in areas that are like fixed blocks, but on diskette they are called *sectors* (see Figure 7-12). Sectors, like blocks, represent the amount of data read from or written to a diskette at one time. A sector might hold one record or several records, depending on the size of the record. A typical sector size is 256 bytes.

FILE ORGANIZATION

Records in a file are physically arranged, or organized, so that they can be accessed *sequentially* or *directly* (by key field). Because of mechanical limitations, magnetic-tape files can be accessed only sequentially. Magnetic-disk and floppy-disk files can be organized so they can be accessed either sequentially or directly.

SEQUENTIAL ORGANIZATION

In a sequential file, records are physically located next to each other on the recording medium (see Figure 7-13). All of the storage space occupied by a

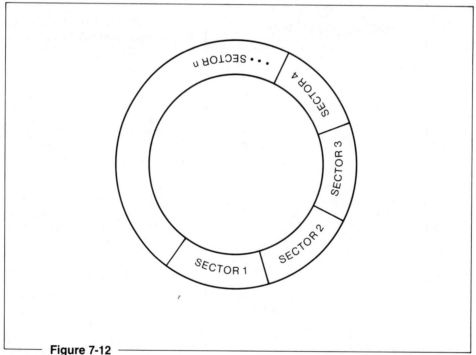

Figure 7-12

Diskette sectors.

sequential file contains actual data records: there is little *overhead* data stored. This is in contrast to files that are organized so they can be accessed directly.

Locating a specific record in a sequential file is relatively slow because all the records stored in front of the one that is sought must be read first. Thus, locating the 10,000th record on the file means reading the first 9999 records.

In order to update a sequential file, we must create a new version of the file—one that incorporates the changes. This includes adding new records to the file, deleting unwanted ones, and changing the contents of data fields on records already in the file. If an enterprise owns several files containing redundant data (recall the customer's name and address stored in each of the bank's departmental files) then, in order to update the redundant field(s), a new version must be made of every file. And to further compound problems, if data in one file is dependent on data in another file, then all file activity must be carefully coordinated to allow access to necessary fields.

INDEXED SEQUENTIAL ORGANIZATION

In a file in which records can be accessed directly, the data records must contain unique key fields. The records are not necessarily physically arranged in key sequence on the storage medium, but they are logically arranged in key sequence. Remember that these files can be stored only on disk or diskette. And in addi-

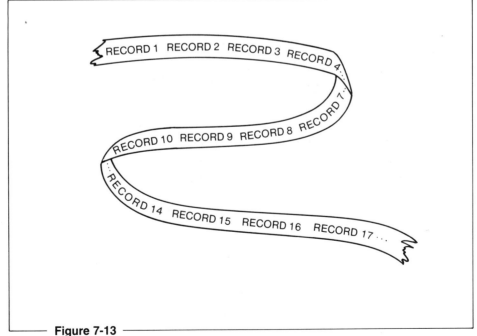

Figure 7-13
Sequential records.

tion to the data records themselves, we must also account for some amount of *overhead* data on the file and unoccupied storage locations. It is some of this overhead data that accounts for the fact that more sequential records can fit on a track than records organized for direct access.

The overhead data allows the computer to get a quick fix on the probable physical location of a specific record. Consider a dictionary: at the top of each page, one line is used to tell the reader the first and last words to be found on the page. The top line on the page then is overhead. That line could be used to store more of the dictionary's text, making the whole book shorter (though it would make it more time-consuming to locate a word). By skimming the tops of pages, we quickly find the page on which our target word is likely to be found.

Indexed sequential files are organized in roughly the same way as a dictionary. In addition to the actual data records, special data called *indexes* are stored that indicate the highest key field to be found on each track and cylinder of data in the file (see Figure 7-14). These indexes allow the accessing program to zero in on the probable physical location of a particular record without having to read all the records in the file. In other words, it can locate a specific record directly.

For example, suppose that a program is seeking a data record whose key field contains the value 0148. It first searches the cylinder index for the first key value that is not less than the desired key value. The first index of 0119 is rejected as too small, but the next value of 0194 is larger than the desired value, so the program proceeds to the indicated cylinder, cylinder 21, where it

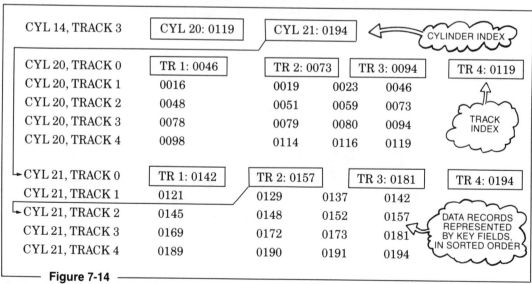

Figure 7-14

Schematic diagram of an indexed sequential file.

searches the track index in track 0 in the same fashion. This time, the key value 0157 halts the search because it is not less than the desired value of 0148, so the program proceeds to search the indicated track, track 2. Here it finds the desired data record, after searching only three tracks instead of the six tracks that would have been searched in a sequential search of the file. If the desired key value is not found in the searched data track, the program knows that it is not in the file.

The amount of extra storage space we must allocate for overhead data depends on the device and the file-access software provided by the computer manufacturer or software vendor.

DIRECT ORGANIZATION

Direct files are organized so that the unique key field for each record in the file provides the basis for its ultimate physical location in the file. Direct files, like indexed sequential files, can be stored only on direct-access storage devices, such as disks. Each record is stored at a unique location, at an address made up of three numbers: the number of the cylinder, the number of the track on that cylinder, and the number of the record on that track.

Like indexed sequential files, direct files allow us to access a specific record very rapidly. And, like indexed sequential files, they require more space than that needed for the actual data records. This is primarily due to the fact that all the locations needed to store data records are not necessarily contiguous. Thus many locations are unoccupied, because no record's key happened to produce the addresses of those locations.

Changes made to records in either a direct file or an indexed sequential file are made in place: a new version of an updated record physically replaces the old version. The application program does not disturb any records that are not updated, because it reads only the record that is to be changed. Records with new key fields are easily added to the file, either by inserting them into an appropriate location or by logically linking them to related records (the linking data are also overhead) while physically storing them in a special area for added records. Records are not physically deleted from an indexed sequential or direct file. Rather, they are *flagged* to indicate that they are no longer active records. This is usually done by including a flag field in each data record in the file and setting the value to indicate whether the record is active or inactive.

FILE ORGANIZATION: A SUMMARY

In summary, sequentially organized files can be accessed only sequentially. Virtually all of the storage space available is used to store data records. Locating a specific record can be time-consuming. Updating the file always means creating a new copy of it, but the old copy can be saved for backup.

Indexed sequential files contain data records that have unique *key* fields, as well as overhead data called *indexes*. The indexes allow rapid access to specific records. Direct files also contain data records that have unique key fields. In order to determine the location of a record in a direct file, we use an algorithm on the key field to produce a physical address. Like indexed sequential files, direct files allow rapid access to specific records.

Indexed sequential files and direct files can be accessed either sequentially or directly. Much available storage space is used for overhead data (such as indexes) or is set aside for future record additions. Updating specific records is a relatively fast process because the target record can be located, changed, and replaced with the new version of the record. Backup copies of the file must be made periodically.

DATA-ACCESS METHODS

Files can be accessed in one of two ways: sequentially or directly. *Sequential* access means that each record in a file is read and processed in order, from the beginning of the file to the end (or to the point at which some other condition specified by the program has been satisfied). Thus, in order to process the 37th record in a file, one must first process the preceding 36 records.

Direct access means that each record in a file can be read and processed without having to read and process all of the preceding records in the file. Thus, the 37th record can be retrieved and processed directly, while the previous 36 records remain undisturbed.

Sequential access is appropriate for some applications, whereas direct access is appropriate for others.

Deciding between sequential and direct access depends on the application in which the data will be used.

If rapid access to specific records is required, then direct access is better than sequential.

If the file is very active (most of the records are processed each run), then sequential access may be a good choice over direct access. On the other hand, sequential access is inefficient for low-activity files.

If data are received in random order, then indexed sequential or direct files may be more appropriate than sequential files.

A highly volatile file (many additions and deletions) organized for direct access will result in many records being added to the overflow areas, slowing down subsequent access to them, and will leave many deleted (flagged) records in the data-storage area. Therefore, the file will have to be reorganized frequently to put all the records into their proper locations. This can be time consuming. On the other hand, sequential file additions and deletions, if done in batches, can reduce the overall time needed to perform file updates in volatile files.

When making data-storage decisions, one consideration is the cost of the equipment. Microcomputer systems generally use diskette, magnetic tape (usually cassettes rather than reel-to-reel), and sometimes a scaled-down version of magnetic disk. All of these options are less expensive than magnetic disk, not only in terms of the cost of the storage medium itself, but also in terms of the cost of the drive that the pack or reel is mounted on. A diskette drive might cost a few hundred dollars, while a magnetic-disk drive might cost several thousand dollars.

Tape is generally less expensive than disk or diskette, but the application may require direct access to specific records and thus rule out tape as an option.

Another consideration is the number of records to be stored. If the number of records is small, then tape or diskette may be appropriate. Of course, disk can be used whether you are storing hundreds, thousands, or millions of records.

It is impossible to say that one device is better than another, or that one access method is superior. Cost and application requirements guide the designer to appropriate choices.

After carefully studying the application, the devices available for storing data, and the various file organizations and access methods, the designer and user must make another major decision: choosing between (1) *building* software to create files and to retrieve, update, and manipulate data, or (2) *buying* either a file-management or a database-management system to do all those things and more.

Because of the cost and other ramifications of using commercial software, the decision to do so is made very early in the development of the system, usually during the second step of system development (considering alternatives to solve the user's problem). Of course, both the user and the system analyst participate in this decision.

FILE-MANAGEMENT SYSTEMS

File-management systems are available from many software vendors, and for a wide variety of computers, ranging from small personal computers to large mainframes. This commercial software allows the user to define the formats of various data files, create them, and then retrieve data from them. Purchasing a file-management system relieves the organization of the time consuming and expensive task of writing its own programs to create and process data files. When the files are created, they can be processed by programs written in-house. Thus, the organization can buy some of its software (file-management system) and build some of it too (application programs).

DATABASE-MANAGEMENT SYSTEMS

Like file-management systems, database-management systems (DBMSs) are available from many software vendors and for a wide variety of computers. Some DBMSs cost less than $1000 and require under 64K of computer memory; others cost more than $500,000 and require several hundred thousand bytes of main memory. They are usually far more complex and powerful than any file-management systems.

Two problems with data storage that are addressed by database-management systems, but not by file-management systems, are data redundancy and data interdependence. We saw above that redundant data is difficult to update, so file maintenance is often handled improperly, incompletely, or inconsistently. We also noted that, if processing data in one file depends on the contents of another file, then file coordination becomes complex, especially from the application programmer's viewpoint. Database-management systems relieve the programmer of many of these problems. Of course, as with everything else in data processing, there are tradeoffs. We will examine them here.

First, two definitions. A *database* is an integrated collection of data. It includes the actual stored data, data about the stored data (overhead), and descriptions of the relationships among data records. Unlike systems in which separate departments maintain their own separate files (often introducing much data redundancy), a database is a centralized repository of an enterprise's data. For the most part, a fact is recorded in only one place: thus all applications that need access to the fact have access to it.

A *database-management system* is a collection of programs and subprograms that handle input and output functions. The stored data files that the DBMS handles are collectively called the database. They are usually stored as indexed sequential or direct files. See Figure 7-15.

ADVANTAGES OF DBMS

Database-management systems provide services in at least two areas: (1) they ensure integrity within a collection of data, and (2) they relieve the application programmer of the need to know physical file details.

INTEGRITY   When accessing and processing data in one file requires access to data in another file, and when redundant data must be updated, the software accessing the related files must be fairly complex. It must account for many variables. When the software to do this is written in-house, the application programmer expends a lot of effort on the process of coordinating file accesses. In a database-management system, the relationships among pieces of data are established when the database is created, and the DBMS software takes care of it from then on. Data redundancy is kept to a minimum, because separate files for separate applications are no longer maintained.

PHYSICAL FILE DETAILS   When using ordinary file systems, the application programmer must be familiar with the physical structure of a data file, the size of each record, the location and size of each data field, the number of records in a block, the type of storage device on which the file is stored, the unique name of the file, and a host of other details. All those details must be described either within the program or within the JCL (job control language) that accompanies the program. We saw in previous chapters that it is the physical data-storage details that change, causing program maintenance problems.

In DBMSs, the application programmer is shielded from all of those details because a DBMS is a subprogram (or several subprograms) that handles all access to the database—including reads, writes, updates, and deletions. Essentially, the DBMS includes the low-level modules that handle file I/O on a pro-

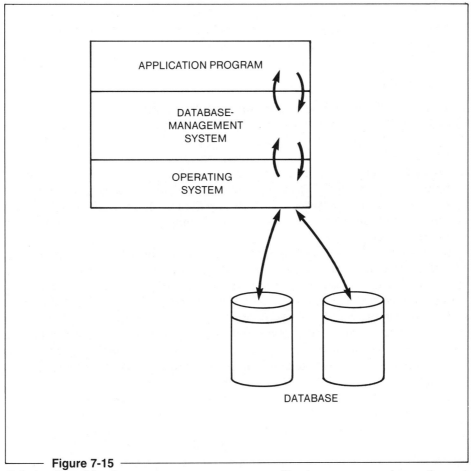

Figure 7-15

Relationship of database-management system to application program and operating system program.

gram structure chart: because those modules are the only ones that require file details, the application programmer does not include them in any other modules. In fact, the application programmer can often ignore physical file details altogether, because the Database Administrator (DBA) or database programmer takes care of them within the DBMS.

COSTS ASSOCIATED WITH DBMS

Do not be misled by the word "subprogram." A DBMS can consist of hundreds of thousands of lines of code, many times larger than the application program calling it, and can cost thousands, even tens of thousands, of dollars! Choosing to use a DBMS incurs many costs: the price of the software, the price of increasing the computer's main memory, the cost of larger and faster disk-storage units, the cost of changing the company's procedures for accessing and updating data (remember, everyone has access to a centralized collection, as opposed to fragmented data dispersed among several departments), the cost of training users, programmers, and operators, and so on.

WHY DATABASE?

Then why would we choose a DBMS rather than a file system? If the organization's data is both extensive *and* needs to be highly integrated, then redundancy and lack of integrity are going to be problems. We choose a database-management system if the application requires rapid access to a large, integrated collection of data. Most of the time it is less expensive to use commercial software than it is to build our own software. The more complex the application's data accesses, the more sophisticated (and ultimately more expensive) the accessing programs must be. In many cases, it is simply cheaper to buy than it is to build.

However, for a DBMS to work, it must know what the data relationships are. Therefore, the designer must first *logically* define the database. This chapter opened with a description of stored data and of relationships among pieces of stored data. Before we can do physical database design, we must first establish and document the relationships among pieces of the users' data. In fact, we illustrate these findings with diagrams like the ones we saw in Figure 7-7. Remember, a database is a centralized collection of the organization's data; *many* users must be consulted, and many different user requirements must be satisfied.

Physical database design requires the skills of an experienced Database Administrator. Not only must the DBA be familiar with the logical data relationships and dependencies, but he must also know the specific commercial DBMS software being used, and the physical devices on which the database is stored. The DBA's job is to determine the physical organization of database files, to create the database, and to be responsible for its integrity and security.

Although physical database design is beyond the scope of this text, Kroenke*
is an excellent source of material on it.

DATABASE: A SUMMARY

Database processing is becoming more popular and more widespread as users
see the need for centralized, integrated data. At the same time, prices of database-management systems are dropping, and computer memory and storage
devices are becoming cheaper. This makes DBMS more financially accessible,
even for the small user. It makes little sense to become bogged down in the
details of file description and manipulation when commercial software will take
care of those things. On the other hand, some applications simply do not require
the sophistication that DBMS provides at current prices.

In order to create a database, we must first determine the relationships
among pieces of the user's data. These relationships are incorporated into a
logical database design. The database administrator (DBA) uses the logical
model to design the *physical* database. Along with the logical model, the DBA
must also be familiar with the DBMS being used and with the physical storage
devices that will hold the database files.

Eventually, all the data are stored on physical devices such as disk, and
usually are organized as indexed sequential or direct files. Of course, databases
occupy more disk-storage space than ordinary files, because they incorporate
much overhead data describing record relationships that are not built into
ordinary file systems.

SUMMARY

Data are recorded facts and figures about people, things, and events. We store
data so we can refer to them again in the future. In computer systems, we often
store data on magnetic storage devices, such as tape, disk, and diskette. When
we store data for future reference, we try to do it in such a way that accessing
them will be fast and easy, while keeping the cost of both hardware and software
in line with the user's needs.

Three popular ways to organize stored data are sequential files, indexed
sequential files, and direct files. Sequential files make most efficient use of
storage space, but they allow only sequential access to records. Indexed sequential and direct files allow very rapid access to specific records, although they
require more storage space on the disk.

The choice of file organization and physical devices is influenced by a number
of application-dependent variables. How quickly must the data be retrieved?

*Kroenke, D., *Database Processing (Second Edition)*. SRA, 1983.

How much data is being stored? How frequently will the data be referenced, changed, added, or deleted? What are the relationships among various pieces of data? The answers to all of these questions affect data-storage decisions.

The physical devices are chosen to meet the user's needs: magnetic tape, disk, and diskette can all be used to store sequential files, whereas only disk and diskette are able to support indexed sequential or direct files.

Software needed to create, update, and access stored data can be written in-house, or it can be purchased from commercial software vendors. File-management systems and database-management systems are available for a variety of computers, and they suit a variety of applications. The choice, and thus the ultimate cost, depends on the user's needs.

A database is an integrated collection of data. A database-management system (DBMS) is a collection of programs and subprograms that create and access the database. Database-management systems are usually more sophisticated, more powerful, and more expensive than software needed for an ordinary file system. Both database-management systems and file-management systems reduce the application programmer's need to know physical file details.

KEY WORDS

Activity   The percentage of records accessed in a file during a single run.

Block   The amount of data transferred at one time from tape or disk into the central processor.

Database   An integrated collection of data, including application data, record relationships, and data about data.

Database-management system   A collection of programs and subprograms that read, write, and maintain a database; also known as DBMS.

Data-storage device   Computer hardware on which data can be recorded and stored, and from which data can be retrieved.

Direct access   A data access method by which a particular data record can be retrieved from a file. See also *sequential access*.

Direct file organization   A method of storing records in a file by which a record's address is derived from the value in the record's key field.

Fixed-size blocks   A file-blocking technique in which file blocksize is determined by the system and is the same for all files.

Flag field   A field in a record whose value indicates the record's current status. It is often used to mark a record for subsequent physical deletion from an indexed sequential or direct file.

Index   Part of an indexed sequential file. The cylinder index indicates the highest key-field value associated with each cylinder of the file, and the track index indicates the highest key-field value associated with each track of data on a particular cylinder.

Indexed sequential file organization   A method of storing records in a file so that the cylinder and track location of a particular record can be determined by searching indices.

Logical database design   Identifying and documenting relationships among data.

Overhead data   Data stored within a file that is not application data but is used by the system to facilitate file access and management.

Physical database design Mapping a logical database design into a particular DBMS format. This activity is usually done by a database administrator.

Redundant data Duplicate data; the same data stored in more than one record or file.

Sector A fixed area on a diskette.

Sequential access A data access method in which records are read or written one after the other, according to their physical position within a file.

Track A circular recording area on the surface of a disk or diskette.

Unblocked file A file in which there is only one application record stored in each block.

Volatility The percentage of records added to and records deleted from a file during a single run.

QUESTIONS AND EXERCISES

1. There are 10,000 records on a file. Once each week, 85% of the file is updated. Would you organize the file as a sequential file, an indexed sequential file, or a direct file? Why did you choose that organization?

2. The application is a central clearing house for sporting-event and concert tickets at a variety of theatres and arenas. Ticket orders can be mailed in, can be telephoned in, or can be placed in person. Phone customers and customers who order tickets in person want to know what seats are available and want immediate confirmation when they place their orders. How would you organize the stored data? Why?

3. In the application of Exercise 2, if you used indexed sequential or direct file organization, what would you use as a key field? Suppose this ticket outlet deals with 25 arenas and sells tickets for events up to three months in advance. How would you determine the number of records in your file(s)?

4. Draw a diagram illustrating the relationships among objects in a school administration system. Students take course from teachers. Teachers teach several courses each semester, usually to several sections. Teachers advise students. Students who live on campus live in one of the six dormitories. Students and teachers take books out of the library. Teachers place books on reserve.

5. What additional information do you need to know about the Agency for the Blind before you can decide how to store the agency's data?

CHAPTER 8

Capturing and Presenting Data

When you complete this chapter, you will be able to—

- design source documents
- design screens for online data entry
- design screens for displaying output
- design reports to be printed on stock paper
- design reports to be printed on custom forms

INTRODUCTION

In this chapter, we examine ways of capturing input data for storage, and ways of presenting output data to someone in printed or displayed format. Because people both provide input and use output, we must keep their needs and abilities in mind when designing forms and reports.

INPUT DATA

Input data are the raw facts coming into our system. Our goal in designing input data is to make the task of entering the data as simple, logical, and error free as possible.

When we approach this task, we design *source documents* on which input data are manually written, and we select the *media* used to enter the data.

SOURCE DOCUMENTS

Source documents are usually pieces of paper on which a user writes data that are subsequently entered into the system from punched cards or from diskette, or even are entered directly into the computer. When designing a system, you may find that some source documents already are in use. Each of the currently used documents must be examined and considered for—

continued use in the new system,

modification for use in the new system,

replacement by another document, or

elimination, if no longer needed in the new system.

If the documents already in use capture data needed by the new system, and if they are easy to complete and interpret, then they can be retained. The designer should evaluate each current source document objectively, and should carefully weigh its merits in the new system before deciding its fate. In other words, an existing document may or may not be retained in the new system.

At least two different people deal with source documents: the user (who might be a client of the company or an employee in a department in the company) and the person who actually enters the data. When we design source documents, we should keep both audiences in mind, and we should design a document that is both easy to complete and easy to transcribe from.

FORM COMPLETION   A source document should be understandable and logical. All areas to be completed should be unambiguously identified. We cannot assume that the user knows what to write: we must tell the user what to write.

A field as simple as date of birth should not be difficult to complete. Yet, unless you make it clear that you expect two digits for the month, two digits for the day, and four digits for the year (written MM DD YYYY), you are bound to get misinterpretation. Europeans, for example, write the day first, then the month.

Sections of the form should be cohesive. For example, collect name and address and identifying information in one area if you can. If there is a limit to the number of characters you allow, then guide the user by drawing a box for each allowed character.

For those sections in which the user chooses from several options, explicitly direct checking of the appropriate box rather than entering a character. See Figure 8-1 for an example of this. Shading can be used to highlight very important areas. Keep the form uncluttered, though. Too many graphics can be distracting and confusing. Simple is better.

Examine the form in Figure 8-2. It is easy to read and easy to complete. The customers who complete this form at home have no trouble interpreting it. It is also easy to transcribe from.

INPUT MEDIA

We can get data into the system in several ways—for example, using punched cards, diskette data recorders, or online data entry. Let us look at how the input medium affects source-document design.

PUNCHED CARDS   Punched cards are typically either 80 columns or 96 columns wide. There is therefore a strict limit to the amount of data that can

CLASS (check one)
- ☐ Freshman
- ☐ Sophomore
- ☐ Junior
- ☐ Senior
- ☐ Graduate

SEX (check one)
- ☐ Female
- ☐ Male

SERVICES REQUESTED (check appropriate categories)
- ☐ Change report format
- ☐ Change report contents
- ☐ Correct processing error
- ☐ Initiate project study

Figure 8-1

Source document with check-off boxes.

be encoded on a single card. The arrangement of fields on a punched card should be both logical and cohesive. Keypunch operators copy data from source documents onto cards, so the activities of designing source documents and card layouts must be coordinated.

Frequently, only some of the data from a source document are actually keyed onto cards. A customer who sends a check to pay his bill, for example, might enclose the top half of the bill he received. This contains his name, address, date of billing, account number, and other data. The keypunch operator might key only the customer's account number and the amount of the check. Therefore, those two areas on the source document should stand out, to make the keypunch operator's job as easy and as error free as possible.

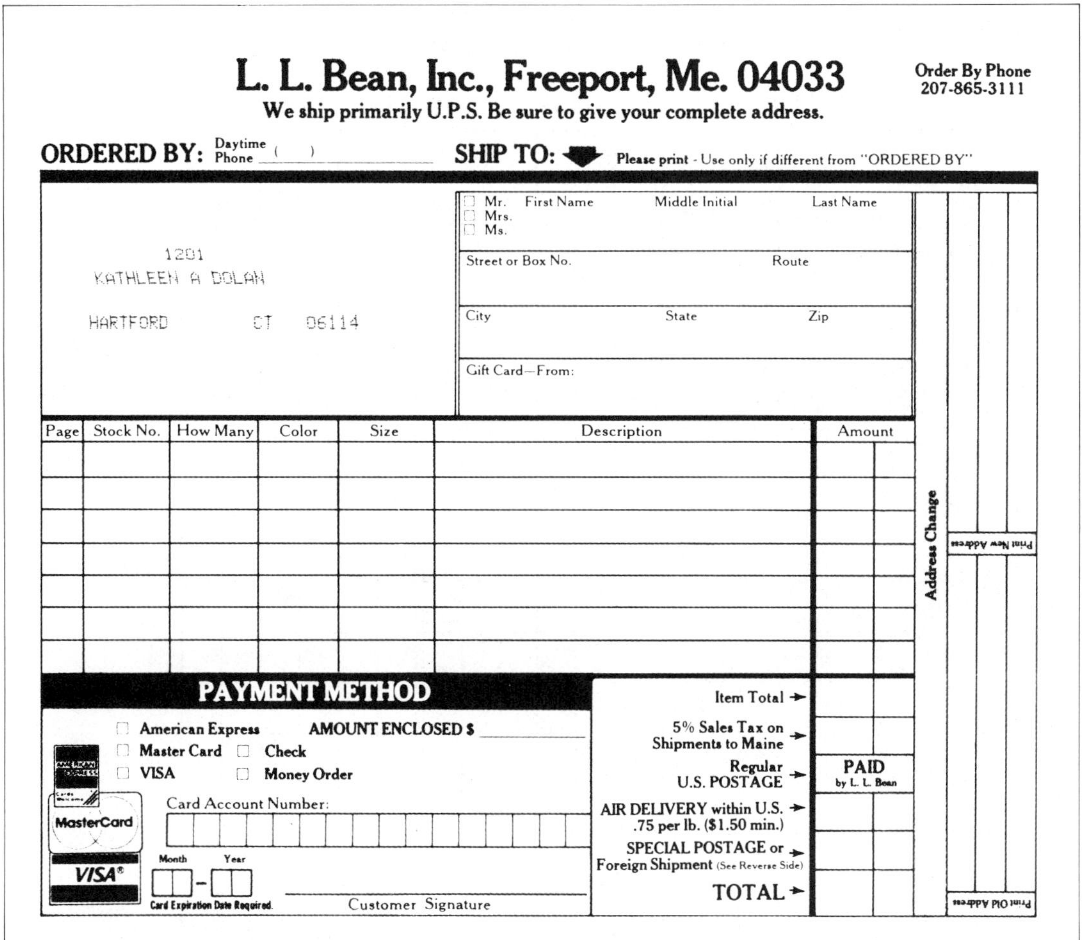

Figure 8-2
Well-designed mail-order form.

Other source documents are designed to represent a *batch* of input cards. In Figure 8-3, each line on the source document is keyed onto a single punched card. Each card will contain a customer's account number and a payment amount. Each card is assigned a unique sequence number: this number is preprinted on the form. The bottom line on the page is different from all the rest of the lines. It is called a *batch control card*. Instead of containing application data, it contains numbers representing the sums of the numbers in the columns above. Although these control numbers are not apparently meaningful (the sum of customer account numbers does not represent anything), they will be punched into a card to serve a useful purpose.

Punched cards can be dropped, lost, or damaged; duplicate cards can accidentally be placed in a deck. The control numbers help the system to detect

		DAILY RECEIVABLES INPUT FORM		
		ACCOUNT # PAYMENT AMOUNT DATE		
T	01			
T	02			
T	03			
T	04			
T	05			
T	06			
T	07			
T	08			
T	09			
T	10			
T	11			
T	12			
T	13			
T	14			
T	15			
C	XX			

Figure 8-3
Source document for batch of records.

these types of errors. The individual who prepares the source document calculates the control numbers. When the data is finally punched and read into the computer system, an *editing program* also calculates the control numbers by adding up account numbers and payment amounts as each record is read. If the computer's control numbers do not match the ones on the batch control card, the batch of records is rejected, and the punched cards are compared to the source document to find the missing or extra cards. Thus, we control the input data going into the system.

KEY TO DISKETTE   Many key-to-diskette systems were modeled after key-to-card systems. They replace clumsy cards with easy-to-handle diskettes, and they increase the productivity of the key operator by automatically performing many special functions. Still, vestiges of keypunch systems linger: records typically are limited to 80 or 128 characters. As with cards, records stored on diskettes are usually processed in batches. In fact, when you design source documents and record layouts for key-to-diskette systems, you should use the same guidelines as for key-to-card input. Make the source document easy to complete and logically arranged, and make it easy for the key operator to follow the form when keying. The operator should not have to skip all around in order to identify the next field to be keyed.

ONLINE DATA ENTRY   As terminals become less expensive, and as personal computers become more and more popular, interactive data entry is being more widely used. There are so many applications for online data entry now that an exhaustive list would be too long to include here. Some popular ones include online library systems, automatic bank teller machines, and hospital patient-data collection.

Online data entry involves a processor that accepts data and commands from a user, analyzes the input, and responds to it by either requesting further input or rejecting the input and helping the user correct it. The input device is usually a keyboard, although other devices such as touch-sensitive screens and voice input can be used as well. There is also a video display device on which data and instructions for the user are written, and sometimes there is audio output as well.

Some keyboards are elaborate full keyboards with all alphabetics, digits, special characters, and function keys; others consist of only a few special-function keys and digits. Some video display devices can display 24 lines of 80 characters all at once; others display only one line at a time on a tiny screen. The choice of hardware depends on the application (doesn't everything?). Regardless of the devices chosen, we must make the task of entering input data as easy and error free for the user as possible.

There are at least three common techniques for getting the user to enter the desired data: menus, prompts, and templates.

THE MENU   The user of an online data entry system frequently chooses from among several available options, just as a diner in a restaurant has a choice of several different entrees. We have adopted the term *menu* in data processing to describe the list of options from which the user of an online data entry system can select.

Figure 8-4 shows a series of menu choices that are displayed to the user of an online student-registration system (this is only one small part of the data captured at registration time: only the part for which menus are appropriate).

The possible entries to each of the categories (class, school, campus, etc.) are both known and limited. Therefore, menus can be used. The person who is using the system need only position the marker (shown in Figure 8-4 as a small box) in front of a menu choice to indicate a selection. Positioning the *cursor* (which shows the current typing position on a screen) might be done by pressing a key on the keyboard, by moving a joystick, or by moving a light pen.

By using menus, we limit the user's possible responses. Therefore we also limit chances for errors. In the student-registration menu, the user need only position a marker and then enter the choice (usually by depressing a RETURN key or an EXECUTE key on the keyboard). You can also design the menu so the user must enter the *number* of a choice (see Figure 8-5). Remember, the user in this case can enter an invalid selection. You must edit the user's responses and allow correction of errors.

THE PROMPT   Another technique for getting the user to enter needed data is by *prompt*ing the user. The system presents a request for data, and the user responds by entering appropriate data. In Figure 8-6, the capital letters represent prompts from the system; lowercase letters indicate responses entered by the user.

In a prompting dialogue, one question is presented, one response entered, then another question, another answer, and so forth. Control passes from the system to the user and back again. Frequently the computer edits some of the data as it is entered. If the data is not valid, then the computer can prompt for corrected data before continuing.

We frequently prompt the user to answer yes or no to some question that will determine the computer's next set of instructions. A question like "DO YOU WANT TO TRY AGAIN?" is common. Phrase the question so the user understands what is expected, and accept appropriate responses. One way to do this is to accept either Y or y as an affirmative response, and any other entry as a negative response. Do not require the user to key y–e–s. Also remember that the general population does not attach the same significance to binary 1s and 0s as data processors frequently do: therefore, avoid them.

THE TEMPLATE   A third technique for guiding the user to enter data is the template. A template is really just an electronic preprinted form on which the user "fills in the blanks." The entire form, or a cohesive portion of it, is presented at one time. The cursor should be positioned at the first "blank" so the

```
        Class
■       Freshman

____    Sophomore

____    Junior

____    Senior

____    Continuing Education Student (nonmatriculating)

____    Graduate Student


        School
■       Arts and Sciences

____    Nursing

____    Graduate

____    New Resources


        Campus
■       New Rochelle

____    Bronx

____    Manhattan


        On-Campus Housing
■       Brescia Hall

____    Ursula Hall

____    Maura Hall

____    Angela Hall
```

Figure 8-4

Data entry by menu for student-registration system.

user can immediately begin to enter the requested data. Whenever it is possible, automatically position the cursor at the next field, relieving the user of the responsibility to move the cursor himself.

Figure 8-7 shows a template as it is displayed on a screen, before any data has been entered. One advantage to using a template is that it can be designed so that is closely resembles the source document, making data transcription easy.

You can save the user some time and keystrokes by placing default values in appropriate fields. In Figure 8-7, the date is keyed onto the first screen only. After that, the same date appears on subsequent screens. If it must be changed, then it can easily be rekeyed. Otherwise, it can be entered without rekeying.

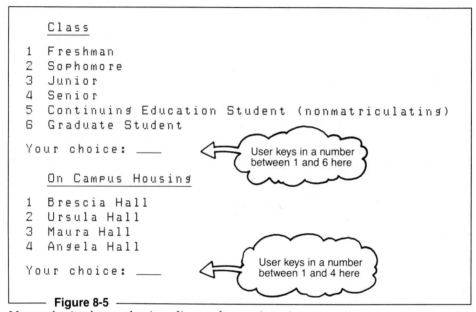

Figure 8-5

Menu selection by number in online student-registration system.

```
ENTER THE FOLLOWING DATA
NAME: stuart little
IDENTIFICATION CODE: d1045
ADVISOR: mary fulbright
STREET ADDRESS: 256 k street
CITY: micropolis
STATE: ct
```

Figure 8-6

Data entry by prompting the user.

If the template is too large to fit on a single screen, you can separate it into *pages,* broken on logical groups of data, or you may be able to *scroll* the template.

If you break the template into pages, then common sense should be your guide to where logical divisions come. It makes sense, for example, to capture data related to the customer's name, address, phone number, and shipping address on one page, and to capture data related to the customer's order on another page. Another page might request information about billing procedures and shipping arrangements.

Scrolling is what movie producers do to show the credits at the end of a movie: lines showing names and contributions to the movie start at the bottom of the screen and migrate upward until they fall off the top. We can do the same thing with templates that do not fit completely on a screen. Simply move part of the text up on the screen, and display new lines at the bottom.

If you decide to scroll, then it is helpful to display the bottom four lines or so of the first screen at the top of the second screen, and display new text on the remaining lines. This allows the user to see where she left off on the previous screen, thus lending some continuity to the data-entry process.

Some designers spend a lot of time making their screen displays artistic or pretty, drawing little boxes around some areas, underlining other areas, or separating lines of text with asterisks. Their intentions are good: they want to point out certain important screen areas to the user. Unfortunately, the screen becomes cluttered, and cluttered screens are difficult and frustrating to read.

The best way to separate areas on a screen is with blank space. For example, if you want to emphasize the fact that customer name and address are different

```
CUSTOMER NAME:              DATE:      07/14/85
STREET:
CITY:               STATE:          ZIP:

SHIP TO ADDRESS:
STREET:
CITY:               STATE:          ZIP:

CAT#      QTY   SIZE   COLOR     DESCR        PRC
_____    ___   ____   _____                 _____
_____    ___   ____   _____                 _____
_____    ___   ____   _____                 _____
_____    ___   ____   _____                 _____
_____    ___   ____   _____                 _____
_____    ___   ____   _____                 _____
_____    ___   ____   _____                 _____
```

Figure 8-7

Template.

from the shipping address, then leave blank lines between the two address areas, not a line of dashes or a string of asterisks. Blank space is much easier on the eyes. Easy data entry will be more accurate data entry.

The general rule for designing screen displays—whether they are menus, prompts, or templates—is to strive for simplicity. The simpler it is for the user to understand what the system expects next, the more likely it is that the user will provide the correct input. Here are some specific suggestions that will help you design good screens.

GUIDELINES FOR SCREEN DESIGN

DO NOT SLOW THE USER DOWN   When we use either prompts or menus to direct the user to enter data, we usually ask for one field or response, then edit it, then ask for another field, edit it, and so forth. This approach can severely encumber the experienced user. We should allow options for the experienced individual to enter data into the system quickly, without having to wait for menus and prompts to appear.

One technique for doing this is simply to accept strings of transmissions from the keyboard, store them in a buffer, and read them one at a time when they are needed.

For example, in a menu-driven accounting system, the user who keys in daily receipts might always use menu item 3 on the master menu (3 means Accounts Receivable), and menu item 6 on the A/R menu (6 means Daily Receipts). When the user signs on to the system, he should be able to enter 3 6 and expect to get to work, rather than having to wait for the master menu, then enter 3, then wait for the A/R menu, then enter 6, then get the system going.

This is also true of prompts. One of the advantages of prompting is that we can guide inexperienced users through the most complex of functions, because we can be slow and deliberate. That is also one of the disadvantages to prompting: the experienced user can enter data much faster than prompting allows. Rather than prompt the user for every individual field, allow the user to enter data in parameter format—that is, a lot of fields entered as a string.

Common applications for keying parameters rather than responding to prompts include bank transactions and airline reservations. Consider a library system in which the user is trying to determine whether certain books are currently in a particular library served by the system. In order to find out, the user must enter a personal identification code, then the ISBN code of the book in question, then the code for the library. This could be done with prompts:

YOUR ID#: user enters id number (pause)

CATALOG #: user enters catalog number (pause)

LIBRARY LOCATION CODE: user keys in two-character code (pause)

(response)

book name IS AVAILABLE AT library name

But the user ought to be able to enter all at once:

identification #/catalog #/library location code (pause)

(response)

book name IS AVAILABLE AT library name

The point is that, if the user is able to work rapidly, the computer should not force a slow pace.

Of course, you may be able to mix both techniques in your system. Determine which of the system features will be used most frequently. If the users will become very adept at entering a few fields at a time, then design the procedure so it defaults to parameters (for example, with airline reservations), but *allows* the novice user to be prompted by requesting help. For less common features, use prompting.

DISPLAYING MESSAGES   In general, messages that are displayed for the benefit of the user should be worded carefully and formatted consistently. They should be clear and unambiguous.

Information messages report on the status of the system. In a word-processing system, for example, an information message might state the name of the document being edited, and the page, line, and position of the cursor. The information line might look like this:

```
document  DESIGN  page  14  line  52  position 71
```

Information within this line keeps changing, of course, as the user uses the system.

Other messages that report on the status of the system can be presented to the user whenever a function is invoked that takes a relatively long time to complete. Suppose the user is making a backup of a disk, and this function is going to take a while. Users become uncomfortable if they are facing a blank screen. They tend to wonder if the system really is working or if something has unexpectedly gone wrong. To set the user's mind at ease (and to prevent some rash action, such as pushing the reset button), display a message that the copy function has been invoked and the system is working. Tell the user what track you are up to, or display a message that the function takes about so many minutes, along with a timer that ticks off the minutes for the user. Whatever you do, don't leave the user with a blank screen.

Error messages seem to be much more difficult to design than system-status messages. Our perspective of the system as designers is so different from the user's. Yet error messages constitute a large part of the human-machine interface, and more program code seems to be devoted to error processing than to normal processing. You would think that by now we would have error messages all figured out. Here are some things you can do to design meaningful and helpful error messages.

Location. When designing screen displays, designate one area of the screen to be used exclusively for error messages, and another to be used to display system-status messages. The rest of the screen can be used for application data, menus, templates, and so forth. This way the user will always know where to look for messages.

Message Numbers. Do not display only error-message numbers for the user. The user has no desire to look up E6392 in a manual to find out what the problem is. Display INCORRECT MENU CHOICE, if that is what is wrong.

What Happened? An error message should state the precise problem. Accompanying prompts and menus will help the user correct what went wrong, but the error message itself should identify the problem. Choose your words carefully. The message INVALID DATA FIELD is not really helpful after the user has filled in a whole template with data. INVALID CUSTOMER NUMBER is a lot more helpful, and *highlighting* the incorrect field draws the user's eyes right to it. The experienced user will look at it and think "Oops, it's supposed to be numeric," and go back and fix it. The inexperienced user should be able to ask for help in order to receive further information on the error. Perhaps you could indicate the editing criteria for the invalid field, like this:

```
CUSTOMER NUMBER, INTEGER, BETWEEN 1000 AND 8599
```

You can draw the user's attention to parts of the screen in several ways. *Highlighting* means displaying characters that appear brighter than all the other ones. *Blinking* means alternately displaying and erasing characters (the cursor typically blinks). *Reverse video* means displaying dark characters on a light background, instead of the standard light characters on a dark background.

Consistent Wording. Choose your words carefully, then use them throughout the entire system. Too frequently we use several different nouns to describe the same item, or several different adjectives to describe the same condition. Consider a message that informs the user that the unit-price field is not correct, that it is supposed to be numeric. One application that accepts and validates unit price issues the message "UNIT PRICE NOT NUMERIC." Another program that validates the field, issues this message: "U PRC NOT NUM," while still another warns: "PRICE INVALID." All of them should issue exactly the same message text. A very useful extension of this technique is also to display some indication of *which module* detected and reported the error—this can help in tracing program bugs.

One of the advantages to using the program-design techniques described in this book is that a designer would undoubtedly have defined a module called "Validate Unit Price" used by all three applications. Because the module generates only one error message, the user would always see the same wording, regardless of the application that encountered the invalid field.

Correcting the Error. Once we have detected an error and informed the user that something is wrong, what do we do? Abort the system? Close down shop and go home? Of course not. We have to help the user correct the mistake and continue.

There are some errors for which no direction is necessary, only a statement of the problem. Suppose, for example, that a user at a terminal tries to get a report printout. We display a message that says: "PRINTER NOT POWERED ON." If the user has control of the printer, it is not necessary to follow this with "SO POWER THE PRINTER ON." In other cases, we need to be more explicit. For example, if the printer is at a remote site, we might conclude the message with "CALL EXTENSION 517."

Our task is to tell the user exactly what to do to fix the mistake. By using a combination of menus and prompts, we can guide the user through the steps needed to correct the error. Let's look at an illustration.

Suppose a user tries to create a data set and store it on a backup diskette, but there is already a data set there by that name. The error message states the problem clearly and unambiguously:

```
DATA SET xxxxx ALREADY EXISTS
```

We can offer the user this menu of error correction options:

```
___ Delete existing data set
___ Add new data set to end of existing data set
___ Rename new data set: _____
___ Cancel
```

Notice that, if the user chooses to rename the new data set, he is prompted to type in the new name.

Erase the Message. Once the problem has been corrected, erase the error message. Resist the temptation to display a cute message such as "THAT'S BETTER" or "THANK YOU."

Audible Tones. Many computer systems are capable of generating audible tones: a click, a buzz, or a beep. Unless the error is one that would otherwise go undetected, do not draw attention to your user with an audible tone. There are several reasons for avoiding audible tones.

Most of us know when we make a mistake, even without the computer.

Most of us do not want to announce our errors to the world (the world being everybody within earshot). We just want to fix them and go on.

Beeps and bells are annoying.

Avoid designing tones into your system. *After* observing users on the job, insert tones only where they are absolutely necessary.

PRESENTING OUTPUT

A wide variety of people read the output from our system: customers read invoices, insurance agents read policy and premium reports, stock pickers read picking lists, store clerks read inventory reports, chief executive officers read sales summaries, government agents read tax reports, and so forth.

We must design our output so it can be quickly read and interpreted. In this section, we examine two popular forms of computer output: printed reports and video displays.

PRINTED OUTPUT

Printers and paper come in all shapes and sizes. Some printers are tiny, fitting a hand-held pocket calculator; others are large, such as the ones used in businesses that produce a large volume of printed output. Some printers make marks on the paper by shooting tiny jets of ink at it, others strike a die against an inked ribbon, while still others use laser beams. Plotters draw graphs and diagrams with multicolored pens. Some printers accept single sheets of paper, others use perforated continuous paper, and still others use rolls of paper. Despite the vast differences among printers, they all have one thing in common: they produce images on paper that a human being is going to read and interpret. When designing output, we are less concerned with the device itself and more concerned with the way data are formatted and presented.

We look here at two general categories of printed output: internal reports printed on stock paper, and output printed on custom forms.

INTERNAL REPORTS   Perforated stock paper is usually barred, or striped, to make it easy for the reader to follow one line all the way across the sheet of paper. Such paper is so popular for computer output that many people refer to it simply as "computer paper." Each line typically accommodates 132 print positions; many printers can print either 6 or 8 lines per inch (lpi), resulting in 66 or 88 printed lines per page.

When planning a report, we use a print-layout chart as an aid (Figure 8-8). Each box on the grid represents a single print position. We start by marking off an area with the same dimensions as the paper we plan to print on. Then we format the report keeping the following things in mind.

Identify the report. This includes the report title, the date the report was printed, and the effective date of the report. The two dates are not always the same.

Number all pages.

Do not assume the reader knows what the data are in a column. Use unambiguous column headers. In the Agency for the Blind, "DATE" might mean the date a book is due, the copyright date of the book, the date the book is needed by another client, or the date a reservation was made. If there is *any* chance the reader will not know, label the column explicitly.

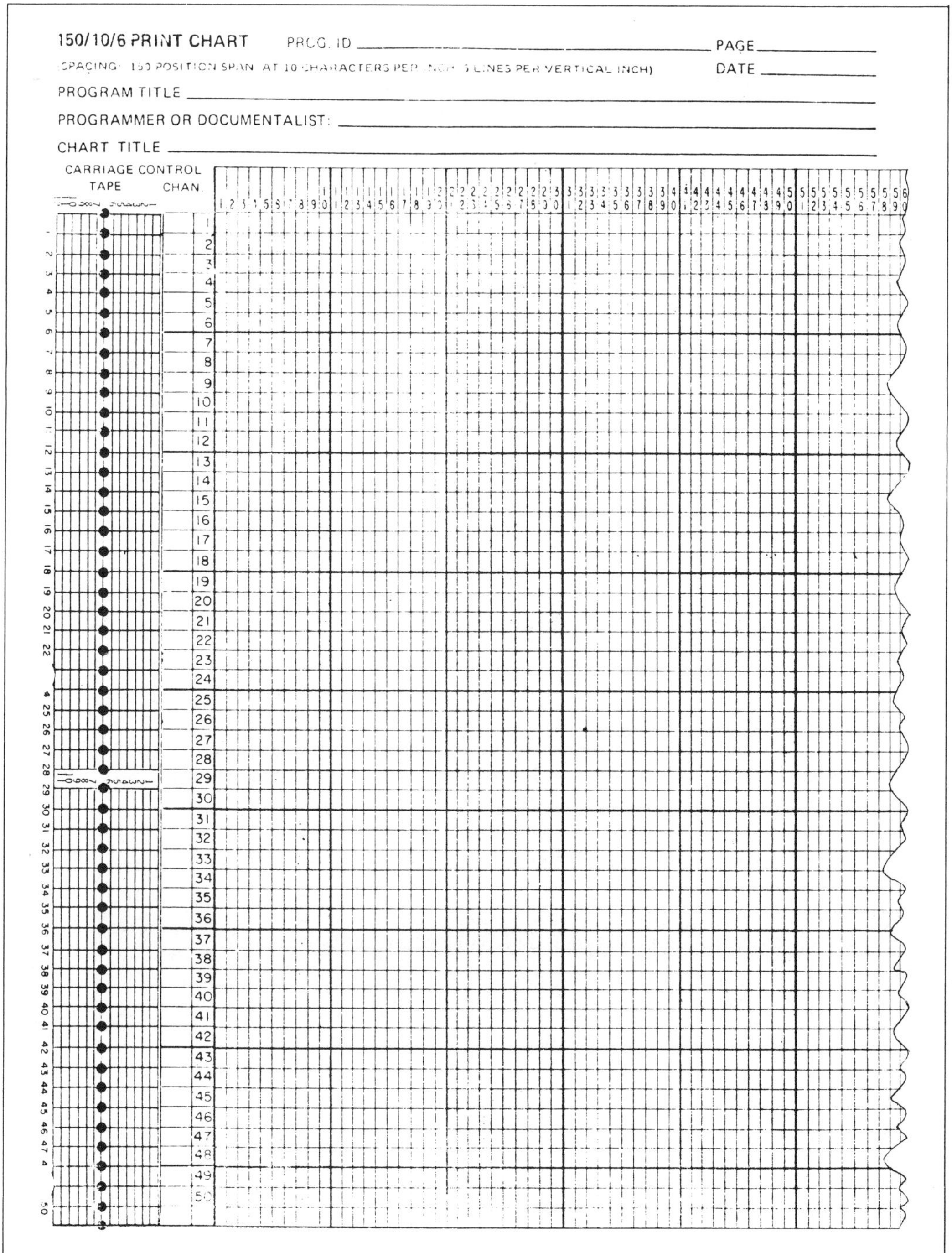

Figure 8-8

Print-layout chart.

Print column headers at the top of every page.

Use commonly accepted data formats. Dates in the United States are in the format MM DD YYYY, whereas in Europe they are printed DD MM YYYY. Do not use Julian dates. Print 5.2% rather than .052 if the user better understands it. Of course, you must know your user. Separate strings of data by embedding spaces. For example, social security numbers are usually written with spaces separating the third and fourth digits, and again between the fifth and sixth digits. It is easier to read 046 43 1378 than it is to read 046431378. Telephone numbers are another example of numbers that are more easily read in groups than in one long string.

Separate lines with blank spaces rather than with asterisks or dashes.

Edit numeric fields: suppress leading zeros, insert decimal points and commas, align columns of figures on the decimal position, and so on. Exceptions to this are identification numbers where even leading zeros are significant (for example, social security numbers, inventory part numbers, bank account numbers).

Identify the end of the report.

CUSTOM FORMS   Custom forms are frequently used for printed output that is sent outside the organization that produces it. Special areas on the form are already marked and identified. Anyone who has ever received a telephone bill, a bank credit-card bill, a school transcript, a mail-order invoice, a bank account statement, an employee W-2 form, or almost anything else produced by a computer, has seen a custom form.

Like source documents (see the section in this chapter on input data), output forms must be designed with the reader in mind. All data areas should be clearly and unambiguously identified. The physical layout should be logical and cohesive.

Custom forms may be *preprinted* (with field identification, named columns, boxes for subtotals and totals, the company name and logo, etc.) or may be printed at the same time as the data (non-impact printers such as laser printers typically produce the "form" on blank paper at the same time they print the output data). Preprinted forms are loaded into the printer when the output is ready to be printed. This physical activity requires the intervention of an operator to change the paper. After the forms are loaded, the operator must carefully adjust them (this is sometimes called *forms alignment*) so everything prints in the correct boxes on the form. The printing program usually contains a special routine that prints constants the operator uses to adjust the printer.

Expert consultation on custom-forms design is available from business-forms manufacturers. They have spent thousands of dollars on years of research on report design. The results cover everything from the use of shading, to printing multiple copies with and without carbon paper, to the best color combinations for reading and for contrast, to dozens of other details that together create an effective form.

If we use stock preprinted forms, such as standard W-2 forms, then we format our data so it is printed in the proper areas on the form. If we design our own forms, then we control both the data formats and the preprinting at the same time. No matter which of these options we face, we should strive for simplicity.

VIDEO DISPLAY

In many online applications, we display data on a screen for the user. This might be one line showing the user's bank-account balance, or thousands of lines listing policy holders for an insurance agent. The worksheet we use looks much like a print-layout chart, but its dimensions match the dimensions of the screen. Screen capacities vary, of course, but some of the standard ones hold 80 characters across each of 24 lines (80 × 24), 80 characters in 12 lines (80 × 12), or 40 characters in 24 lines. Block out the areas reserved for messages and system status information (Figure 8-9). The remaining boxes can be used for displaying data.

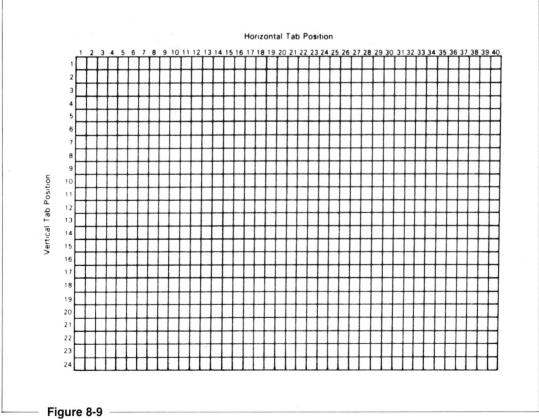

Figure 8-9

CRT screen layout chart.

When designing displayed output, we use the same guidelines we use for designing printed reports and for prompting the user to enter input.

Separate areas with blanks rather than characters. An uncluttered screen is easier to read.

Use unambiguous column headers. If you scroll, then scroll only the data, not the headers: they should remain fixed on the screen.

Signal the end of output data for long lists.

Present data in a format understandable to the reader: edit numerics, divide long strings into smaller groups of letters or digits.

Align numeric fields on the decimal position; left-justify alphanumeric fields.

Use reverse video sparingly. Reverse video can be effective if used only where you must draw attention to a particular area. If used too often or in too many places, it loses its effect.

All in all, strive for simplicity. You will increase your chances of designing readable and understandable output screens.

Your ally in video-display design is the user who will be reading it. Involve the user by preparing some sample screens and asking for feedback. Take the user's criticisms to heart. Modify the screen to suit the user's satisfaction. After all, it is the user who will ultimately judge the success or failure of the system.

SUMMARY

In a computer system we must capture input data for processing and storage, and we must present output for the user to read. Because input data are provided by people and output is read by people, we must keep their needs and limitations in mind when designing source documents, screens, messages, and reports.

Our goal in designing ways to capture input data is to make the process as easy and error free as possible. We design source documents that are easy to interpret, complete, and transcribe from. After the form is filled in, a data-entry operator usually keys data from it onto cards or onto a data recorder. We want to design the form in such a way that it is both easy for the data-entry operator to read it, and easy for a person to complete it.

Online data entry may reduce the need for specially trained data-entry operators, depending instead on the user to enter data. Online data entry requires the use of screens, which must be designed every bit as carefully as source documents. We can use menus, templates, and prompts to get the user to enter data. At the same time, we must edit incoming data, detect errors, and guide the user through error-correction procedures.

When designing output screens and reports, our goal is to present the data so they are easily read and interpreted. We can identify fields and columns, use preprinted forms, separate output areas, edit fields, and make judicious use of special video effects. The guideline to follow in both report and video-display design is: simple is better.

KEY WORDS

Batch A collection of records processed as a group.

Batch control card A punched card containing numeric values calculated manually. Computer instructions perform the same calculations and compare answers to the values on the control card. This is a technique for identifying discrepancies between cards that are processed and cards that should have been processed.

Blinking A technique for alternately displaying and erasing a symbol on a CRT.

Cursor The indicator of current typing position on a CRT screen.

Forms alignment Adjusting the paper in a printer so that output prints in the correct location.

Highlight To intensify the brightness of symbols displayed on a CRT.

Menu A list of options from which user can choose.

Prompt To present a message telling the user what to enter next.

Reverse video A CRT display technique of displaying dark characters on a light background.

Scroll To move lines up a CRT screen, losing lines from the top and adding lines to the bottom.

Source document A document containing a written record of data to be entered into a computer system.

Template A CRT display that presents a "form" into which a person enters data.

Video display A CRT, or screen.

QUESTIONS AND EXERCISES

1. Design the screens for the following online system. Be sure to account for system information and error messages.

   The system will add, multiply, and divide.

   If the user chooses to ADD, he must enter up to four numbers. The sum will be presented. Errors are entering less than two numbers, entering more than four numbers, entering nonnumeric characters.

   If the user chooses to MULTIPLY, he must enter two numbers. The product will be displayed. Errors are entering less than two numbers, entering more than two numbers, entering nonnumerics.

   If the user chooses to DIVIDE, he must enter two numbers: the first number is the dividend, the second one is the divisor. The quotient, up to three decimal positions, will be presented. Errors are entering less than two numbers, entering more than two numbers, entering nonnumerics, entering a divisor of zero.

2. Design the input screen(s) for the order form in Figure 8-2.

3. In order to do some statistical analysis of your class, we need to gather input data from each student, including age, height, hair color, date of most recent head cold, favorite color, and an indication of sensitivity to these allergens: pollen, animal danders, newsprint, laundry detergent. Design the source document that students can fill in. Assume that the data will be punched onto cards.

4. Repeat Exercise 3, but assume that data-entry clerks will enter data from each form at a terminal. Show both the source document and the screen.

5. Repeat Exercise 3, but assume that each student will enter his or her own data at a terminal. Show source document(s) and screen(s).

6. Design a screen that displays the following information for the registrar's office: student social security numbers, names, and major(s), in alphabetical order by class.

7. Design a computer-generated overdue notice for your school's library.

8. Locate three custom forms you have received (examples: phone bill, transcript, W-2 form, bank statement, registration form, meal contract, automobile registration, charge-card statement) and critique them. Look for poorly identified or unidentified fields. Is there any data you cannot understand? Can you readily locate your account number, the date the document was printed, and the date payment is due (if it is a bill)? What changes would you make? Why?

CHAPTER 9

Designing Procedures

When you complete this chapter, you will be able to—

- design user procedures
- design operations procedures
- document procedures in reference-manual format
- document procedures in video-display format

INTRODUCTION

In this chapter, we define procedures and discuss who needs procedures and why they need them. We examine some techniques for analyzing tasks and deciding on the steps to accomplish the tasks. We discuss documentation, both for operators and for users. We examine the organization and layout of reference manuals, and we review screen displays as they relate to procedures. We conclude this chapter by identifying various tasks for which procedures must be established, both for operators and for users.

WHAT ARE PROCEDURES?

A procedure is a series of steps followed in a regular definite order. Procedures, as we discuss them in this chapter, are the steps people must perform in order to use or operate a system. Procedures are steps people follow, just as programs are steps a computer follows. And when people follow the procedures, they are able to control the rest of the system, making it do what they want it to.

There are similarities between procedures and computer programs. Let us examine some characteristics of programs.

They are complete, leaving nothing to chance.

They are precise. Computers are not capable of guessing.

They are unambiguous. Programming syntax does not allow for ambiguity. A term, once defined, can mean only one thing.

They are written in a language the computer understands. This language includes both syntax and format.

They are designed and reviewed before they are coded.

They account for the detection of errors, and error handling.

They direct the computer to *do* something.

The procedures we establish for users and operators must have the same characteristics. They too must be complete, precise, unambiguous, written in a language the reader understands, designed and reviewed before they are cast in stone, and must tell the reader how to detect and correct errors. And they must direct the reader to *do* something. Procedures are meant to answer the reader's question, "What do I have to *do* in order to ...?"

Only after the procedures for performing functions have been designed, reviewed, and approved, are they set down permanently: they are written in manuals, summarized on reference cards, embossed on equipment, or stored for display on video screens. Writing of documentation is discussed in the second part of this chapter. We must decide *what the procedures are* before we can write them down.

DESIGNING PROCEDURES

Designing procedures is not unlike designing computer programs. We first identify the functions we want the operators and users to be able to perform. Then we decompose each function into its many tasks, decompose each task into subtasks, and so forth, until we reach an appropriate level of detail for our audience. We review all the steps and make adjustments if necessary.

Consider the procedure for reserving a book in the Resource Center at the Agency for the Blind. The procedure answers the librarian's question, "What do I have to do in order to reserve a book?" The procedure might be as follows.

> To reserve a book you need—
> the client's name, the teacher's name, the title and author of the book, the client's reading mode, and the date the book is needed.
>
> 1. Verify that the client is eligible. If not, reject the reservation.
>
> 2. Look up the book in the card catalog.
>
> 3. If the book is not in the catalog, reject the reservation.
>   If the book is checked out and won't be back before the date needed, reject the reservation.
>   If the book is not reserved for anyone else, reserve it.
>   If the book is out, but it's due back before the date needed, reserve the book.
>
> 4. If you were able to reserve the book, write out a confirmation for the teacher—otherwise, return the reservation request to the teacher.

Based on the user's familiarity with each of the tasks identified above, further definitions might be needed. For example, a new librarian might need to know how to verify that a client is eligible. Does that mean that the client's name is on file? Does it mean that the client is under a certain age? Enrolled in a certain type of program? Is there a limit to the number of books a client can have out at a time? An experienced librarian might know the answers to all those questions. But procedures are written for everyone who uses the system, experts and novices alike. Thus, another procedure might be required for verifying a client's eligibility, answering the question "What do I have to do to find out if a client is eligible?"

IDENTIFY EACH FUNCTION

Much of the spade work for these activities has been done, if you followed the analysis and design steps described in this text so far. We know many of the functions for which we need procedures, because they were identified during the specification step. For example, in the Agency for the Blind system, we must establish procedures that tell the user what to do in order to maintain the

Client File, maintain the Library Books File, assign a teacher to a client, change a client's funding sources, order a new book in print or in braille, catalog a book, reserve a book, print retrieval reports, and so forth. The computer programs will actually do the file maintenance, record retrieval, and report writing, but we must tell the person how to invoke each function, how to get necessary data in, and how to correct or respond to errors.

Write the name of each function that the user and the operator will need to perform. The names of these functions will be made up of a strong verb and a direct object (just like module names and process bubble names). Some examples are—

Correct batch error
Enter accounts receivable
Get your bank balance
Print customer statements
Reserve library book
Request system change
Backup master file
Initialize diskette

DECOMPOSE EACH FUNCTION

For each function you identified, break it down into smaller tasks, or steps. Some people call this task analysis. In program design, we call it functional decomposition. No matter what the label is, the intent is identical: to take a big problem and break it up into little ones.

For your own purposes, you may want to write the several smaller tasks in pseudocode format; or you may prefer a flowchart, or a decision tree. As long as you are precise, the technique you choose here is irrelevant.

WALK THROUGH THE STEPS YOURSELF

This is particularly helpful for tasks that involve physical motion, such as readying peripheral devices, loading paper into a printer, preparing a batch of cards for input, or keying data at a terminal. Often the sequence of steps is awkward in real life although it looks fine on paper, and it must be modified. Make adjustments as you go along. When you are satisfied with the procedure from your point of view, submit it to the highest authority: the user or operator who will perform it.

REVIEW STEPS WITH USER AND OPERATOR

At this step in system design, it is more vital than ever to call in the users and operators. We must get their input and feedback now: after the system is implemented, it will be too late. Why do we need them? Because users and operators see the system from a different point of view than we do. And it is their point of view that interests us now. We are establishing procedures for them. What seems like a great procedure to designers is sometimes awkward, difficult, or impossible for the user or operator. It is worthless to spend months in system development producing an elegant, powerful, maintainable, expandable, efficient system that the user cannot use and the operator cannot run!

If your system will be installed in-house, then you have ready access to your audience. If not, find someone with roughly the same background as your target audience. Then walk the user or the operator through the procedures. Tell him what to do next, every step of the way. Note points at which he needs further explanation or clarification or both. Ask him for his expert opinion: Do the steps make sense from his point of view? If not, change them. Observe him carefully. Sometimes he will not ask for help: he will simply guess or perform some task incorrectly. You must clarify the procedures to eliminate or at least minimize the user's confusion and anxiety.

AN ILLUSTRATION

Here is an example of a procedure that made perfect sense to the designer, yet seemed ill-designed to the users:

When taking telephone orders for a large mail-order company, the order-entry clerks were directed to interview the caller/customer to first get the name and address, then each item wanted (including quantity ordered, catalog number, size, color, and so forth), and finally to ask for the customer's charge-card number. When entered, the charge-card number would be compared with an online file of known fraudulent accounts and overcharged accounts, and occasionally rejected. When put into practice, the order-entry operators noticed that they frequently entered dozens of line items, only to have the order rejected because of a bad account. They could not believe anyone would be naive enough not to have the charge-account number entered first. They eventually brought this problem to the attention of the system developer, who quickly changed the procedure.

How much easier it would have been to test the procedures with real operators before launching the system. Any order-entry operator could have told the designer where the problem was, if only the designer had asked. The designer had been careful to include the user departments during system-requirements specification, and during database and report design. Unfortunately, the designer did not recognize the value of user input at the procedures-design stage.

You do not have to make the same mistake. Walk the user and the operator through cach procedure. If this involves keying data online, then design your screens and your procedures together. As problems are discovered, adjust both screens and procedures accordingly. Ask the users and operators for suggestions. Then incorporate their suggestions into the procedures. Remember, the user and the operator will not be impressed with loosely coupled, highly cohesive modules; they will be impressed with the ease (or difficulty) with which they can use the system.

DOCUMENTATION

Having decided on the best steps that can be taken to accomplish some activity, we must turn to a related task—that of documenting the procedures for the user and the operator. Documenting means writing the procedures for future reference. Although this might seem easy at first, it is amazing how few people are able to write procedures clearly and precisely, even if they are quite adept at performing them.

The points we will discuss here are the following: know your audience and write for it; respect your reader's experience and limitations; choose the documentation mode that is best for the application; build your documentation directly from the procedures you developed and reviewed with the user and operator.

The objective of documenting is to capture procedures in a format for future reference. The more frequently a person performs a task, the less likely it is that the person will need to look up the procedure in documentation. Of the myriad functions a user or operator might perform throughout the entire system, there will be some that will be done very frequently; others will be options that are performed less frequently; still others will be esoteric functions that are performed very infrequently. It is not surprising to find that 80 percent of the time is spent using only 20 percent of a system's functions.

People have various backgrounds, education levels, work experience, system familiarity, attention spans, anxiety levels, and psychological needs. Although it is not our place to delve deeply into the personality of our readers (both users and operators), it is important to have some understanding of the typical reader. It makes no sense to use in our documentation terminology that a user is not capable of understanding, and thus put the user off. Nor does it make sense to talk down to an experienced system operator. We must know our audience so we can gear the documentation to it. If the typical audience spans a wide spectrum of ages, education, and experience, then create separate documentation geared to each type of user.

How do you get to know your audience? If they work for your company, talk to them. If you have been developing the system with them right along, then you already know them. If your audience is not in-house, then you must have some idea of who will buy and use your system once it is developed. The user audience should have been determined way back in the user-requirements specification stage. Talk to some individuals who are like your prospective audience. Or enlist the aid of professional market surveyors. For a price, they will tell you almost anything you would like to know about your typical user.

The point is that, before you rush into writing reference manuals and designing fancy fold-out reference cards, you must establish

1. what you are going to say (the procedures), and

2. who you are going to say it to (the audience).

Having accomplished both of those vital steps, you have a clear picture of your objective. Now you must decide on the best documentation mode.

PRESENTING DOCUMENTATION

There are two ways we can present documentation: by printing it in hard-copy or manual form, and by storing it in the computer for retrieval and display when it is needed. Regardless of the medium, documentation is used to remind the user of the steps that must be performed in order to complete some procedure. Let us examine the characteristics of written and displayed documentation.

WRITTEN DOCUMENTATION

Written documentation includes reference manuals, reference cards, operations documentation, and instructions embossed or engraved directly on a piece of equipment.

REFERENCE MANUALS   These contain the collected procedures for the entire system. They are usually arranged so the reader can refer to one complete procedure at a time. The table of contents helps the reader locate the needed procedure quickly. Figure 9-1 shows the table of contents from a user reference manual.

One distinct advantage of written documentation is that paper lends itself to illustrations, something that is more difficult to accomplish with screen-displayed documentation. A reference manual that is carefully organized, with the text for a procedure on the lefthand page and accompanying illustrations on the righthand page, can provide clear and comprehensive documentation. Conversely, a reference manual that is organized so the reader must flip pages back and forth to match text with illustrations can be frustrating and confusing.

Figure 9-2 shows pages of a reference manual that illustrate the procedure for changing a printer ribbon.

When planning a reference manual, depend heavily on what you learned when you established the procedures. Collect related procedures under topic headings. Each topic heading might refer to one major system function. In the Agency for the Blind system, the reference manual might have these topic headings:

Resource Center Functions

Client-File Maintenance

Management Reports

Equipment Operation and Maintenance

System Utilities

Thus, the procedure for loading paper into the printer would be described under the section on equipment operation and maintenance; initializing a diskette is described in the section on system utilities; adding a new client to the Agency files is described under client-file maintenance.

Figure 9-1

Table of contents from user reference manual.

Reference to one procedure might be made within another procedure. For example, one of the Resource Center functions is to print retrieval reports. Within the description of the procedure for printing retrieval reports, one of the steps might be "Load forms in printer and power printer on." If the person who is printing retrieval reports is not familiar with the procedure for readying the printer, he or she would look up that procedure under equipment operation and maintenance.

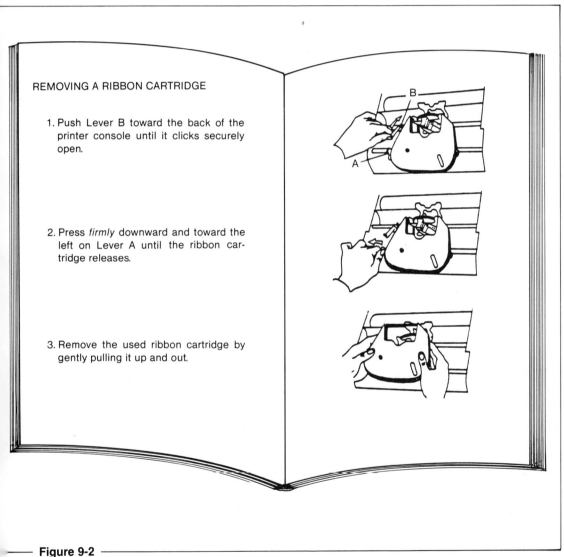

Figure 9-2

e *of illustration in reference manual.*

A reference manual is not complete unless it includes a section describing every error the user might encounter, as well as the procedures for correcting or responding to the error. Often this is done in one section at the back of the reference manual—called "Error Messages," or a similar title. If error messages include message numbers, then organize the section according to message numbers; if not, then organize it in alphabetical order on the text of the message.

It is important that the errors be clearly identified, and that corrective action be stated explicitly. The user will have realized that an error has occurred, and will already have guessed at a solution (or several possible ones). The user will not look up the procedure in a manual until out of ideas, and probably frustrated at the failure to fix the mistake without help. Therefore, it is vital that error-correction procedures be spelled out clearly. Tell the user what to *do* to correct the mistake.

Include every possible error. There are few things more frustrating than encountering an error and not finding any reference to it in the documentation. Even extremely rare errors caused by system or hardware malfunctions should be mentioned, even if the procedure is "Contact your XYZ customer service office." Some developers are reluctant to include pages and pages of error messages and explanations, thinking that it looks bad for the product. How absurd! Every user knows that errors will occur. And every user would rather have access to an explanation and corrective steps than to be left in the dark.

REFERENCE CARDS   Reference cards highlight the steps of each procedure. They are not detailed descriptions of procedures: detailed descriptions are found in the reference manual. Reference cards capture the essence of procedures and serve to jog the reader's memory. They can be kept handy at all times because they are small compared to the reference manual. They are also inexpensive. An installation with many users might distribute reference summaries to each of them, and keep a few reference manuals around for shared use.

Clearly, reference summary cards do not replace reference manuals, but they are very useful. Figure 9-3 shows part of a reference card.

OPERATIONS DOCUMENTATION   In installations with centralized data processing, computer operators are responsible for running all programs. This responsibility includes scheduling programs to be run, ensuring that input files are made available for the programs, labeling and storing output files (such as tape and disk), and bursting and delivering output reports. Few operators are psychic: so just as we are responsible for telling the user the procedures for getting the system to work, so we must provide the computer operator with the directions needed to run the system.

The documentation we provide to the operations staff includes the following:

a list of the programs, and the order in which they must be run;

the identification of all input files to be made available for each program;

identification of all output files produced by each program;

identification of any special forms used by each program;

instructions for the storage and distribution of output;

all messages generated by the programs at the operator's console;

expected response to each message;

error-correction procedures;

special procedures, such as file backup.

A *system flowchart* illustrates file inputs and outputs as well as the sequence of programs that are dependent upon one another (see Figure 9-4).

Forms usually are identified by a form number and a name. A name is sometimes not enough, because several user departments in a company might use forms to which they have given the same name. Although the department

Figure 9-3

Part of a reference card.

262

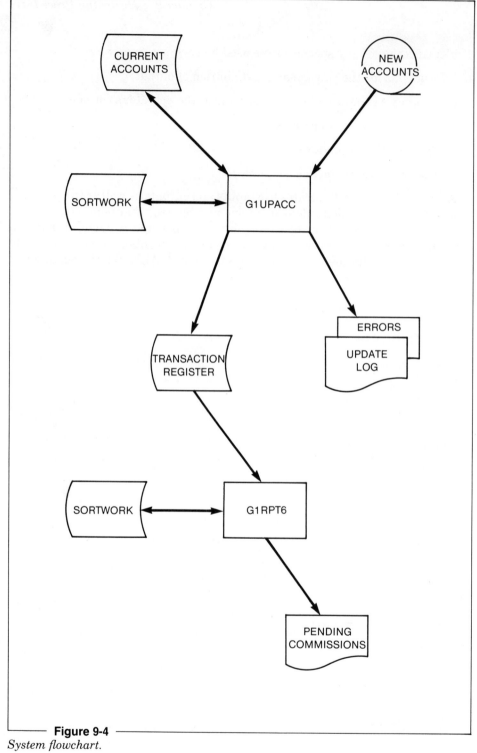

Figure 9-4
System flowchart.

handling individual insurance policies calls its form "customer invoice" and the department handling group insurance policies also calls its output "customer invoice," these are actually two different forms. Custom forms commonly are printed with an identification number in the lower left corner; list the form number as well as the form name so there will be no confusion.

Output reports go to someone who wants to (or has to) read them. That is why they were printed. Make it clear to the operations staff who they should send the reports to if they are responsible for delivery, or who is authorized to pick up the reports from the operations room.

Messages printed or displayed at the operator's console should be designed as carefully as the messages displayed for the user. Messages are used to indicate the system's current status, to prompt the operator to do something, or to inform the operator of an error condition. A program that processes several volumes (reels of tape or disks) might be designed to issue a message telling the operator the number of the volume now being processed. A program that prints output on custom forms might display a message telling the operator to mount the forms on the printer, and to perform the forms-alignment routine. Another program might check the volume and serial number of an input file and display an error message for the operator if the expected file is not available.

In the case of error messages, it is vital not only to state the problem, but also clearly to outline the procedure to rectify the problem. Operations staff, like users, need a list of all error messages they might encounter, and the corresponding procedures. Assume nothing. Tell the operator precisely what to do to correct or respond to the error. If you want to write documentation that is understood, imagine that your reader will be a person encountering the problem at 3:00 A.M., having worked an 18-hour shift the night before, and on the verge of getting the flu. If you write for that person, your documentation will be effective.

Normal procedures are those that the operator performs each time she runs the system. Special procedures are done only occasionally. They must be clearly documented because the less frequently a task is performed, the less likely it is to be remembered.

Operations documentation might be presented in the form of a reference manual, with descriptions of each subsystem collected under the subsystem heading, messages and their explanations in another section, error-identification and correction procedures in another, and special procedures in another section. For an individual program, the sole documentation might consist of a run sheet, like the one illustrated in Figure 9-5. It simply and briefly identifies input and output files, messages and expected responses, special forms needed, and distribution of the output. Whether writing a reference manual or a simple run sheet, the objective is to give the operator all the direction needed to run the job correctly.

INSTRUCTIONS ATTACHED TO EQUIPMENT　We are all familiar with instructions attached to equipment. Open the back of a pocket calculator or a

STEP NAME	INPUT				OUTPUT								
	Data Set Name	Device Type	Vol Ser or Reel #	Dispo-sition	DISK OR TAPE				PRINT				
					Data Set Name	Device Type	Vol Ser or Reel #	Dispo-sition	Printer #	Form			Copies
										STD	Form #		

SUBMITTER _____ DEPT. _____ DATE _____ PAGE _____ OF _____

Console Prompt:

Reply:

Comments:

Figure 9-5
Operator run sheet.

camera flash attachment: there is an illustration that shows how the batteries are supposed to go in. Copying machines frequently have the operating instructions glued to or printed on the surface; they also show how to change paper, clear a jammed paper path, and add toner. All this information is available without a reference manual.

If you develop a system that uses special equipment, you may wish to attach the operating instructions directly to the machinery. Automatic bank-teller machines, for example, have instructions printed right on them.

If your system uses function keys on a keyboard, you can design a template that identifies the function, rather than the preprinted number, of each key the system uses (see Figure 9-6).

DISPLAYED DOCUMENTATION

Like written documentation, displayed documentation guides the reader through the steps of some procedure. The difference is that written documentation is at the reader's disposal in its entirety, whereas displayed documentation is given to the reader only a piece at a time. Why is that? Because the screen on which we display procedures is the same screen the user needs to read a report or to enter data. We cannot leave our instructions on the screen *and* allow the user to enter data or read a displayed report. In addition to limited space on the screen, another consideration is that people have relatively short memories. They cannot be expected to read a screen with seven or eight (or fifteen) steps listed, commit them to memory, and execute them perfectly.

We solve this problem by presenting only one step of a procedure at a time. As the user or operator performs the step, the system analyzes the response, continues if an expected response is obtained, or signals an error if not. Of course, we tell the user or operator how to respond to an error. After successful

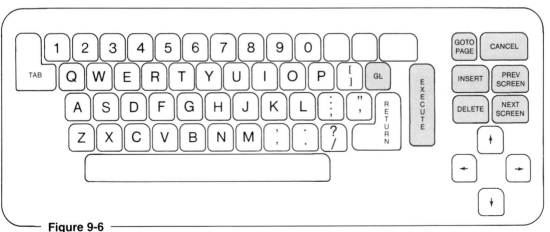

Figure 9-6
Keyboard layout showing function keys (shaded).

completion of one step, we present the next one. This loop—present instruction, analyze response, correct error if necessary, present instruction—continues until the entire procedure is completed.

In Chapter 8, we discussed the use of menus, templates, and prompts to get the user to enter the expected data. Although we presented these techniques as they related to input data, it is natural to consider them to be displayed procedures. After all, we use menus, prompts, and templates to tell the user how to get data into the system, how to correct errors, when and how to activate devices, how to identify herself to the system, and so forth. We lead her through each process, step by step, analyzing each response and guiding her through correction procedures. Clearly, all the topics discussed in Chapter 8 regarding screen display apply here.

The important point to remember is that the procedures must be worked out before you design your screens; no amount of good screen display can make up for poorly designed procedures.

When you design messages to display to the computer operator, be sensitive to the nature of that job. Many designers have never operated a computer, and are therefore totally unaware of what life in a computer room is really like. Because of this shortcoming, it is vital to the successful operation of the system to consult with an experienced computer operator when designing and documenting operating procedures. As a student, you may be far removed from operations. If it is possible, meet some of the computer operators at your school or at a local business. They will give you a completely different, but related, perspective of computer systems.

PROCEDURES THAT REQUIRE DOCUMENTATION

Documenting operations and user procedures is no small task. What follows is a list of some of the procedures that must be included in your system documentation.

Operations
> How to run the system under normal circumstances.
> How to detect and correct errors.
> How to backup files.
> How to recover the system in case of failure.
> How to label and store data sets securely.
> How to identify persons authorized to handle system input and output.
> How to prepare and verify input data (data-entry department).
> How to implement authorized system changes.
> How to maintain records and report on system activity.

User

How to prepare batch input data.
How to submit batch input data.
How to enter data online.
How to detect and correct errors.
How to perform minimal hardware maintenance.
How to request system modifications.
How to activate each system option.

SUMMARY

Procedures are sets of instructions that a person follows in order to use or operate a system. In order to establish procedures, we identify each function, break it down into small steps, review the steps with the user or operator, and modify and adjust them if necessary.

We record procedures so they can be referenced when they are needed. Reference manuals and operator-run instructions contain complete descriptions of procedures. Reference cards contain only brief summaries of procedures, serving to jog the reader's memory. Procedures can also be displayed on screens. Whereas a procedure can be described in its entirety in a reference manual, use of a screen limits the number of steps that can be presented to the reader at one time. Therefore, we are careful to present small tasks within a procedure when guiding a user or operator interactively.

Procedures must be established for users so they know how to prepare and enter data, interpret output, invoke various system functions, and correct errors. Operators must know how to set up and run the system, how to distribute output, how to label and store various files, how to correct errors, how to backup files, and how to keep records on the operation of the system.

KEY WORDS

Documentation A record of the information needed to use a computer system, including procedures for normal operation and for error handling. It can be written in hardcopy form or displayed on a CRT.

Functional decomposition Breaking down a procedure into a series of steps.

Operations documentation Information needed by an operator to run a computer system.

Procedure A series of steps followed in a regular, definite order; steps followed by people in order to use a computer system.

Reference card A summary of important system procedures and other information. The reference card is designed for quick reference only.

Reference manual A written document containing all the information needed to use a computer system. It usually includes procedures for normal operation and error handling, special functions, troubleshooting, and error message summaries and descriptions.

Task analysis See *functional decomposition.*

EXERCISES

1. Locate a reference manual. It can be for a computer system, for a computerized application, or for a non–data-processing application. Some possibilities include the owner's manual for a camera, an automobile, or a tape recorder. Perhaps your college's student handbook describes procedures for various things: registering for and withdrawing from courses, requesting transcripts or letters of recommendation. Select *one* procedure and critique it. Is it complete? Is the text well organized? Do you understand the terminology? Are illustrations useful? Do they match the text? Are you able to follow the procedures written there? How would you change them?

2. Find an interactive system somewhere: a video game, a 24-hour-a-day banking machine, a library book-inquiry system, an online administrative system at your school. Select one procedure and perform it, recording the steps. What were you asked to do each step of the way? If you made a mistake, what was the procedure for correcting the error? Did you encounter anything that presented a problem? How would you improve the procedure?

3. Once each month, the Agency for the Blind updates its Library Books File with records on new books that have been received and prepared for distribution. Write the procedure for the librarian to prepare the input data, submit it to be run against the book file, and correct errors.

4. Write the instructions for a computer operator to update the Library Books File (see Exercise 3). What files are needed? What are some errors that might be encountered? What about correcting errors? How often must the operator backup the Library Books File? Will procedures be different if the file is sequential rather than indexed sequential or direct?

5. Repeat Exercise 3, assuming that the updating is done by the librarian online throughout the month. How does this change your procedures?

6. In the situation described in Exercise 5, how often must the Library Books File be backed up? Assuming there is no computer operator, where would you document the backup procedure?

CHAPTER 10

Designing Training Programs

When you complete this chapter, you will be able to—

- write training objectives
- design tests to determine that objectives were met
- outline a training program
- compare various training presentation media

INTRODUCTION

We train people for only one reason: to change their behavior. We want them to be able to perform required tasks. Who needs training for a new system? The people who will have to interface with the system (operators and users) but do not yet know how. In other words, the people who are not yet familiar with the procedures that have been established (see Chapter 9). Our goal is to design appropriate training programs for the people who need them.

Training takes time and effort on the part of both the trainer and the trainee, time that otherwise could be spent being productive. Therefore we want to train only those individuals who really need it, and we want to do it effectively as well as quickly and inexpensively. When we design training programs we must do at least the following things:

identify the objectives of the training program,

devise a method to determine when an individual has met the objectives,

identify the individuals who need training,

outline the training material,

select appropriate presentation media.

Let us examine each of the steps.

WRITING OBJECTIVES

The analyst and the user established the objectives of the system-development project when they wrote the user requirements. An agreed-to set of user requirements became the ultimate target for all the rest of the activities. When these user requirements are satisfied, the project is complete.

Similarly, our training objectives specify what the trainee must be able to *do* in order to be considered ready to interface with the system. We state our objectives in terms of activities rather than in nebulous terms like "understanding" or "appreciating," because activities are measurable. When the trainee demonstrates that he can do what we want him to do, we let him get to work. It should be no surprise to you that objectives are stated in the same way as modules are named: using a strong active verb and a specific direct object.

First we state what the needed behavior is. Here are some examples:

The trainee will be able to—
retrieve a record.
enter a data field.
initialize a disk.
run a program.
load fan-fold paper.

correct a keying error.
update a file.
burst invoices.

By way of contrast, here are some statements that would not be measurable training objectives, although it would be nice if the trainee happened to develop them along the way:

The trainee will—
understand the system philosophy.
comprehend the updating process.
appreciate the speed of voice input.
trust the system security.
enjoy the job more.

Thus, it should be clear that objectives must be stated in terms of measurable or observable activity.

Second, we state the criteria for acceptable performance, if there are any. This step zeros in on *how well* the trainee must perform in order to be considered trained. In some cases it is not enough that the trainee can do something: she might be required to perform within a time limit or to a certain degree of accuracy. Here are some examples.

The trainee will be able to—
enter 15 telephone orders in less than 30 minutes.
retrieve normal customer-balance information in 30 seconds or less.
correct batch-validation errors in five minutes or less, with an error rate of not more than 1%.
load form #1369 into printer and perform forms-alignment routine in less than 2 minutes.
log-on to system in fewer than four attempts.

It is important, of course, to keep in mind what the trainee will be doing on the job when you are writing training objectives. The individual who is learning how to use a personal computer at home is probably under few time constraints. Conversely, a terminal operator using an online system will be expected to operate very rapidly, and with a very low error rate.

Make your objectives realistic. Remember that you are training human beings. Of course the objectives should reflect those of the system: for example, if speed is paramount, then be sure you train for that; however, do not place undue importance on speed if it is not important during system operation.

TOOLS FOR MEASUREMENT

Having stated your objectives, you must devise a way of determining when a trainee has acquired the target skills. This means you must test performance.

Performance tests must be geared to the stated objectives. The objectives say the trainee will be able to do something; the test must direct him to do that something. Some examples will serve to illustrate this step.

Objective: To enter 15 telephone orders in 30 minutes or less.
Test: Seat trainee at active terminal with a headset, and call in 15 telephone orders from an office phone. Time the session.

Objective: To log-on to system in less than four attempts.
Test: Position trainee at terminal and direct him to log on. Count the number of tries.

Objective: To update four pages of text on a word processor from a marked draft with an error rate of less than 2%.
Test: Give trainee the word-processing document and the marked draft, and direct him to update it. Calculate the error rate.

Be careful not to ask the trainee to do something for which he was not trained. For example, if your objective is that the trainee will be able to create a backup copy of a diskette, then it is not appropriate to test him by asking him to *list the steps* he would follow to create the backup copy; rather, you should direct him to actually create the backup.

IDENTIFY THE TRAINEES

We mentioned above that we do not want to provide training to people who don't need it. The easiest way to weed out the ones who are not in need is to ask the potential trainee to take the test(s) you just developed. Whoever performs satisfactorily does not need training. If that happens to be everybody associated with the system, then you can skip the rest of the training procedures. If someone requires training, then continue with the next step, outlining the training program.

OUTLINE THE TRAINING PROGRAM

In Chapter 9, we saw that the first step in documenting a procedure is to break it down into a series of separate steps. Training programs are designed in much the same way. We decide on the overall goal of our training program, then focus on the skills involved in each of the steps taken to achieve that goal. For example, if the goal of one training program is to enable the librarian in the Agency for the Blind Resource Center to use the online book-retrieval-and-cataloging system, then he must be trained to perform all the various activities: starting the system, powering on equipment, loading paper in the printer, initializing diskettes, labeling and storing backup copies of files, entering data, correcting errors, and so forth. There is so much to cover; where do we start?

We start with the most basic and fundamental procedures. They are not necessarily the easiest ones, but they are the ones without which the trainee cannot function. After laying a foundation of system familiarity, we can build upon it by presenting the more advanced system procedures. Finally, we can present sophisticated system functions. Let's look at fundamental procedures first.

FUNDAMENTAL PROCEDURES

The most basic procedures are those needed to get access to the system. In an online system, the trainee must learn how to log on. The trainee must know how to power-on the equipment and adjust it for comfort. This might be as simple as becoming familiar with a keyboard and video-display terminal. Or it might entail examining a more elaborate configuration including storage devices such as tape and disk or diskette, one or several printers, and alternative input devices such as microphones, graphics tablets, and card readers. Regardless of the complexity of the equipment, the overall goal is the same: the trainee will know *what to do* in order to activate the system.

In a batch system, the trainee might have to learn the procedures for preparing and submitting programs or data to be processed. This includes identifying himself as an authorized user, identifying the program to be run, supplying necessary data, and following the run instructions.

In both batch and online systems, getting access to the system is the most basic procedure the trainee has to learn. The procedures for this step have already been established (see Chapter 9); all we must do now is make sure the trainee knows how to perform them.

After the user has acquired access to the system, he can begin to learn some of the system's basic functions. We start by presenting the user with the most useful functions. There are several reasons for starting here.

If the trainee sees the usefulness of the system early in his training, he will be more favorably disposed toward mastering and using the system effectively.

Even while training is taking place, the user can actually use the system. Thus he will reinforce and perfect his skills, eventually performing useful functions quickly, accurately, and confidently. Furthermore, this enables us to release the system in phases or versions.

If we begin by presenting functions the trainee already knows, then he is learning only a new *procedure*. This makes the training process much easier.

Let us turn to the Agency for the Blind Resource Center for an illustration. You will remember that there are a variety of system functions available to the librarian:

Catalog New Book
Order Book Transcription
Order Book
Inquire Book Status
Check Book Out
Check Book In
Print Retrieval Reports
Reserve Book

Of all the system functions, which would the librarian perceive as most useful? There are two candidates: Catalog New Book, and Inquire Book Status. If cataloging were presented first, then the librarian could actually begin the process of building the library books file by cataloging the existing collection. When the librarian sees how easy the first system function is, he or she will be eager to learn how to use other ones as well.

MORE ADVANCED SYSTEM FUNCTIONS

Now the user feels comfortable with the system and recognizes how it is an improvement over the "old way." We have laid the groundwork for presenting more advanced and less familiar functions. In the Resource Center, these might include some of the system utilities, such as file backup; or they might involve loading custom forms in the printer. Of course, as we present each of the procedures, we also indicate what errors are likely to occur and how to fix them or respond to them.

SOPHISTICATED SYSTEM FUNCTIONS

Finally, we present sophisticated system functions. These are ones that require a solid understanding of the other system functions, or ones that are likely to be invoked infrequently. Often, the training objectives for these functions are simply that the trainee will know how to locate the procedures in the reference manual, and will be able to follow them. We cannot expect the trainee to be able to perform them as quickly and efficiently as the fundamental procedures, such as turning on a terminal.

What we produce, then, is an outline of the training program. It indicates the sequence in which topics will be presented in the training program, the objectives for each training unit, and a means of measuring trainee success. Now let us examine some alternatives for presenting the training material.

SELECT THE PRESENTATION MEDIA

The medium or media you select will depend upon several factors:

the size of the trainee audience,

the homogeneity of the trainees' abilities,

the speed with which skills must be acquired,

the training budget,

the appropriateness of the medium to the task being learned,

the number of times the training is likely to be presented.

You should consider those factors when deciding among presentation media, because some media are more effective than others in certain situations. Your training program either will be presented by a *teacher or trainer,* or will be *self-paced.* A self-paced program is one in which the trainee controls the speed with which the material is covered.

Teachers or trainers are used for two types of training: classroom training, and on-the-job training.

As you would expect, *classroom training* usually includes lectures, discussions, demonstrations, and practice sessions with a group of people. The teacher controls the pace of the presentation. This is an effective approach when the training audience is large, the trainees' skills are similar, and the material being presented lends itself to the classroom setting. It is also appropriate for small groups of people when the material will not be presented very often, or if the content changes frequently.

On-the-job training involves a teacher or trainer, but this time working with one or two individuals at a time. This is truly learning by doing, or learning as you go along. In on-the-job training, the teacher or trainer might simply be someone who is skilled in the use of the system—someone already familiar with the procedures the trainee is learning. On-the-job training is appropriate when the number of trainees is small; when it is *not* critical that the trainee perform quickly, efficiently, and perfectly while learning; and when the training takes place infrequently.

Self-paced training programs focus on the trainee, who may be expected to learn by reading material in a textbook; by listening to taped lectures; by watching videotapes, film strips, or slide programs; or by a combination of these activities. Self-paced training programs are appropriate when the training audience is large but it is not possible to gather together a group at one time (trainees may work different schedules, or turnover may be high); when there are no skilled trainers available; when the number of trainees is very small (for example, only one new hire per quarter); when the material being presented remains relatively constant; and when the material will be presented frequently.

One advantage of self-paced instruction is that the trainee can move quickly through material she finds easy, and slow down to study and repeat material she finds difficult to understand. It is worth noting, however, that many individuals have difficulty motivating themselves to learn material on their own. Therefore, self-paced training programs must be carefully designed to move through the material quickly yet thoroughly, so the trainee accomplishes as much as possible in as short a time as possible.

A written training manual should be carefully organized and formatted. One advantage of printed matter is that illustrations can be included to clarify topics. Pictures are worth a thousand words, but only if they are printed on the same page as or facing page to the text that refers to them. The training manual should be a convenient size and shape. Pages should lie flat when the book is open. For manuals that will be read while the trainee is using a keyboard, the manual should stand up.

Presentations that include slides, video tapes, movies, taped lectures, and combinations of these must be professionally produced if they are to be of high quality. Multimedia programs involve scriptwriters, graphics artists, filmmakers, sound and light engineers, and many other specialists. Your primary contributions to the production of a multimedia presentation are your list of training objectives and the outline that shows the sequence of topic coverage. If you have done this preparation work well, then an expert can guide you through the rest of the lengthy process.

SUMMARY

Training programs are needed when there are individuals who are unable to follow the procedures they need to use the system. In designing a training program, we must state our objectives in measurable terms, devise a way of measuring the trainee's performance, determine who needs training, establish the sequence of topics to be presented, and select an appropriate presentation medium.

Teachers and trainers are used for classroom training and for on-the-job training. Self-paced instruction focuses on the trainee, who is responsible for mastering the material at his or her own pace. Choice of an appropriate approach can be made only after considering the size and skill level of the audience, the speed with which trainees must acquire skills, the amount of money available for training, and the number of times the material will be presented.

Self-paced study materials can include one or more of the following: written training manuals, taped lectures, film strips, video tapes, and slide shows. Professional writers and audio-visual experts have the skills and resources needed to produce quality multimedia training materials.

KEY WORDS

Classroom training   A directed training program that takes place in a traditional classroom setting and usually includes a teacher or trainer.

Objective   A statement of what a trainee will be able to do upon the completion of training program.

On-the-job training   A training pro-gram that is conducted in the trainee's workplace, usually under the direction of an individual who has the skills the trainee is trying to acquire.

Performance criterion   A statement of how well a trainee will perform upon completion of a training program.

Performance test   A test administered

to the trainee in order to determine whether or not the stated performance objectives have been attained within the stated performance criteria.

Presentation medium Materials used to present the training program.

Self-paced training program A training program in which the trainee controls the speed with which the material is presented.

Trainee An individual being trained.

EXERCISES

Select an activity that you know well (wrapping gifts, changing a tire, finding a square root, taking close-up photographs), and do the following exercises. When you finish, you will have designed a training program.

1. Identify the objectives of your training program. Be sure to state them in terms of behavior. What will the student be able to do after completing the training program?

2. Devise a test to determine if the objectives have been met. Identify performance criteria, if there are any (e.g., will be able to change a tire in less than 10 minutes).

3. Identify your trainee audience. Do they have any special needs?

4. Outline your training program. Which topic will be presented first, which one second, etc.?

5. Select a presentation medium.

6. Assume that your training program will be presented to thousands of people nationwide. Will you use the same medium as you selected in Exercise 5? What one, if any, will be more appropriate?

PART 3

After Design

CHAPTER 11

Implementation: An Overview of the Next Step

When you finish this chapter, you will be able to—

- explain how an implementation plan is developed
- explain the importance of reviews of progress against anticipated deadlines
- describe how each system component is built and separately tested
- create module test data
- explain the purpose of system testing
- describe various types of system test cases
- describe three approaches to acceptance testing

INTRODUCTION

In this chapter we discuss implementation, the final step in system development. It is during implementation that we actually build, test, and integrate all the system components we designed in the previous steps. Full treatment of implementation warrants a text of its own, so this chapter is meant to be an overview. However, we also hope that, in seeing an overview of the implementation step, you will have a greater appreciation for the importance of good design.

Implementation includes developing an implementation plan, building and unit testing each system component, integrating all the components into a system and testing it as a whole, then releasing the finished system for production. After the system is in production, user feedback is sought so we can plan system upgrades, add new features, or introduce new technology as it becomes available.

THE IMPLEMENTATION PLAN

The implementation plan is developed by a senior member of the development team or by the manager responsible for the system being developed. The plan includes

identification of personnel and their responsibilities;

calendar schedules and delivery dates for all products;

acceptance criteria for all products;

procedures for error detection, correction, and recordkeeping;

contingency plans for disasters.

Identification of personnel and their responsibilities simply means assigning work units to programmers, analysts, users, technicians, trainers, technical writers, data-entry personnel, and so forth.

Calendar schedules and product due-dates are based on knowledge of the number and skill levels of the personnel available, the number and complexity of programs to be built in-house, the delivery and installation dates of hardware, the availability of development resources (such as computer time), the amount of data to be converted or created, and so on. Orchestration of many individuals and activities both inside and outside the organization that is developing the system requires skill and patience. Planning is absolutely crucial to the success of this step.

Often critical dates are identified, dates that mark the delivery of some product without which subsequent steps cannot be undertaken. An example might be the delivery of a voice-input device, without which the testing of voice-input program modules cannot be done.

Acceptance criteria are the minimum standards a product must meet in order to be considered complete.

A vital part of the implementation plan is sufficient time for normal error detection and correction, as well as a contingency plan to be put into action in the event of a disaster. Only the most naive planner assumes that there will be no errors during implementation. All the good analysis and design tools known to system developers will not prevent real errors from sneaking into real code, or prevent the head of the data-entry department from contracting the flu on the day that file conversions are supposed to begin.

These normal problems should be anticipated, and procedures for handling them should be part of the implementation plan. This means, of course, that when the plan is being developed, *ample time should be scheduled for error detection and correction.* Procedures also should be established for keeping records on the number and types of errors detected: this information will help in making estimates during the next project.

Disasters are thankfully few and far between—fires that destroy equipment and files, labor-union strikes, whole development teams that become ill or quit in the middle of a project—yet there should be a contingency plan in place in the unlikely event that disaster occurs. Regularly backing-up files and program libraries, and then storing them off-site is an easy and relatively inexpensive way to turn a potential disaster into a minor inconvenience. Making arrangements to borrow or lease equipment if yours is destroyed is useful, but only if files and libraries can be recreated from backups.

Whole development teams do occasionnally leave projects in midstream. In anticipation of this, data-processing management should require that systems-development documents (DFDs, structure charts, data dictionaries, module specs, test data, documentation outlines, training schedules, etc.) be kept on file. A company should not depend on individuals to keep system-development information in their heads: they generally take their heads with them when they quit, walk out, become ill, or die. In the event of a disaster, another team can use the existing documents for background and, in a relatively short time, can pick up where the others left off.

The implementation plan is based on deadlines and the assumption that those deadlines will be met. Frequent reviews of progress relative to those dates will result in the detection of problems near the time they occur (as opposed to six months later), and while there is still a chance that the schedule can be salvaged by directing more effort at the problem point. In spite of herculean efforts, some deadlines will not be met. If estimates have been made honestly and one step is late, then all subsequent steps will also be late. Schedules should be adjusted accordingly, and all affected personnel should be notified of the revised schedule.

IMPLEMENTING EACH COMPONENT

We now look at each component, examining *who* builds and tests it as well as *how* each group does its job.

HARDWARE

Depending on the requirements of the new system, we may have to

upgrade existing equipment, making it faster or increasing its storage capacity—for example, adding computer memory;

replace one piece of equipment with another one—for example, replacing tape drives with disk drives;

install computer equipment for the first time—for example, install a word processor, a microcomputer, or a mainframe.

Sometimes we must adjust the physical environment in which the computer equipment is going to be operating. Electronic and mechanical equipment is sometimes sensitive to high or low temperatures, static electricity, radio transmissions, or changes in electrical power. Some equipment requires special electrical wiring; some large computers are placed on raised floors so the cables connecting pieces of equipment can be run under the floor. Adjustable, comfortable office furniture may be needed for the people who will use some pieces of equipment. These might include adjustable supportive chairs, low work surfaces for keyboards, and nonglare overhead lighting. Hardware manufacturers supply specifications for adapting an environment for their equipment.

In addition to environmental controls (air conditioning, antistatic carpeting, special electrical wiring), we must also consider the security of the equipment. We should protect our computer hardware from damage due to fire, flood, theft, power outages, unauthorized use, and many other things. Sprinkler systems and halon gas help control fire in a computer room. The location of the room itself can help protect equipment from flood damage or sabotage. Backup generators can be installed that automatically prevent sudden power drops from affecting the computer equipment, and other regulators can protect the equipment from damage due to sudden power surges.

Access to the computer equipment can, and often must, be controlled. Some security systems allow access only to those people who can identify themselves as authorized personnel. This may be done via special keys, by entering a password or identifier to release a door or turnstile, or by matching voiceprints, fingerprints, or signatures. Many companies hire security guards to help protect the premises.

In addition to the centralized computer equipment mentioned above, some equipment may be stored in a user's work area. Securing it may be as simple as bolting it to a work surface, moving small pieces into locked storage areas at the end of the work day, or securing the entire area. The amount of money and effort invested in security systems usually depends upon the value of the equipment being secured. Thus security varies widely from one installation to another.

Who is responsible for the installation and testing of hardware? Once again, this depends upon the system being installed. Some equipment can be installed by anyone able to follow the manufacturer's directions. Other equipment is too sophisticated or complex for the average business person, and requires the skills of a trained technician. Computer-equipment manufacturers often send their own service technicians or recommend hardware-service companies to install the equipment.

Before the equipment is delivered, the site may have to be prepared. This could require the coordinated efforts of electricians, carpenters, glaziers, air-conditioning professionals, carpet layers, even plumbers. Scheduling the various activities requires good managerial skills.

Equipment is usually installed in one of three areas:

in a central computer room,

in the user's department or work area,

in an area accessible to the public, such as a shopping mall, an airport terminal, or a public library.

Each of the locations requires different security measures to protect both the actual equipment and access to the computer system.

After the area is prepared and the equipment is installed, it must be tested. Testing equipment usually involves executing a planned sequence of tests, called *diagnostics,* designed to reveal hardware malfunctions. Professional hardware technicians know what tests to run, either through experience or because the tests are supplied by the manufacturer. Owners of personal computers or small business computers simply follow the testing procedures supplied by the manufacturer.

Ideally, new equipment has been installed and tested before program testing begins (we'll look at program implementation in the next section), allowing for the most realistic tests. However, if the equipment cannot be available by the time program testing begins, program development can be approached in at least three ways: (1) programs can be written that simulate the new equipment as closely as possible; (2) test time can be purchased from another organization that has the equipment we are waiting for (that is, we test our programs at another company and pay them for the use of their equipment); or (3) we can delay the coding and testing of modules that interface with the new equipment until it arrives, and direct our first efforts toward modules not so dependent on physical equipment.

In any event, new equipment must be installed in time to test the integrated system. It is also very helpful, although sometimes not crucial, to have the actual equipment in place by the time we start training the users and operators.

PROGRAMS

In this section we address two broad categories of programs: those that are purchased (called commercial software), and those that are built and tested in-house. The major difference lies in the fact that commercial software, at least theoretically, is already coded and tested; therefore it is ready to be used "off the shelf." Programs we build in-house take much longer to implement. They must be coded, compiled, tested, debugged, retested, and so forth until they finally work (or until they behave acceptably enough). We will first look at commercial software, and examine how and when it is selected, as well as who is involved in the selection process. Then we will turn to building our own programs: who does it, when it is done, and how we test the programs.

COMMERCIAL SOFTWARE   System software—such as telecommunications systems, database-management systems, operating systems, language translators, and many others—is almost always purchased. Commercially produced application software can also be found in abundance: programs that handle financial applications, inventory, airline reservations, school administration, agriculture, mail order, business planning, and literally thousands of other applications.

Very often it costs less to buy a commercial software package than it does to develop our own programs. Commercial software can carry price tags ranging from less than one hundred dollars to hundreds of thousands of dollars. It is expensive to buy programs only if it would cost you less (in time and money) to build your own.

After the user's requirements are established, the system analyst proposes alternative approaches to solving the user's problems. One or several of the options usually include the purchase of commercial software. Depending on the application, there may be a complete commercial application package that can be purchased and installed in a very short time, thus eliminating the need to design any programs at all! In any event, the decision to buy software vs build it is made very early in system development, and involves both the user and the data-processing department.

There are publications that contain descriptions of commercial software, usually arranged by application. Thus a company interested in inventory control would be able to look up brief descriptions of commercial programs that deal with that topic. More detailed information is always available from the software vendor. Careful comparison shopping with regard to price, hardware required, capacity, speed, performance, reliability, upgradability, and other important factors, can result in the purchase of the most appropriate software packages for both system functions and business applications.

We mentioned that the *selection* of commercial software is done after the user's requirements have been established. Commercial software should be *purchased and installed* by the time it is needed for testing in-house programs, assuming there are some. Often commercial software cannot be installed until hardware changes have been made. Planning is required to coordinate the delivery and installation of equipment, and the delivery and installation of commercial software.

PROGRAMS BUILT IN-HOUSE Let us now look at how we build and test our own programs in-house. For many years, this step has been done exclusively by specially trained computer programmers. The widespread use of computers in the home, in schools, and in businesses—coupled with the development of easy-to-use programming languages—has made writing programs possible even for "nonprogrammers." Accountants, homeowners, engineers, school teachers, insurance salespeople—almost anyone willing and able to learn a programming language—can write programs. There are some approaches to programming that can make it both easy and rewarding.

CODING First, we use the structure chart as the guide to implementation. We have designed a system made up of highly cohesive, loosely coupled modules; controlling modules are found near the top of the structure chart, whereas physical worker modules are located at the bottom. One way to approach programming is to work from the top down.

Starting with the highest module on the structure chart, we code, test, and debug it until it functions according to its module specifications. Then we code one of its workers, test it along with the boss to make sure the combination works, and debug it if it doesn't. Now we add another worker, test the new combination, debug it, add another new module, and continue until all the modules work with each other. Figure 11-1 shows a possible sequence for coding and testing the modules on a small structure chart.

You might have noticed that, when a boss module is coded and tested, it will fail if there are no worker modules for it to call. In COBOL, for example, all PERFORM statements must refer to a paragraph within the program; in BASIC, all GOSUB statements must refer to line numbers within the program; in COBOL, PL/I, and assembler language, all CALL statements must refer to a separately compiled subroutine that has been linked to the calling program. In other words, in order to test any calling module, we must write all its worker modules. Does this mean that we must write the entire set of interfacing modules before we ever test them? Not at all. Instead of writing complete versions of worker modules, we can write incomplete modules, called *stubs*.

Program stubs can be nothing more than empty modules—modules that return control to the boss as soon as they are invoked. For example, we can code a COBOL paragraph name without any instructions in the paragraph. Or we can write a subprogram in which the only executable instruction is GOBACK.

Program stubs are more useful if they do something: they might display a message indicating that they were entered; they might return a constant to the calling program; they might perform simple calculations, or simplified versions of their actual calculations; they might test their input parameters. In fact, they can do some of these things in combinations. The purpose of writing programs this way is that we can quickly code and test modules. Once a module is working properly, we no longer pay attention to it, concentrating our efforts instead on the modules yet to be developed. This rational, organized approach to building programs is possible only because we did such a careful job of designing first. Now we use our design as the guide to building and testing the modules.

Naturally, code reviews are used throughout the programming step, just as design reviews were used throughout design (see Chapter 6). The objective is the same: to find errors as early as possible. The sooner we find errors, the more quickly and easily we can correct them.

Many installations have *programming standards*—rules for coding to which programmers are expected to adhere. Standards are established so the company

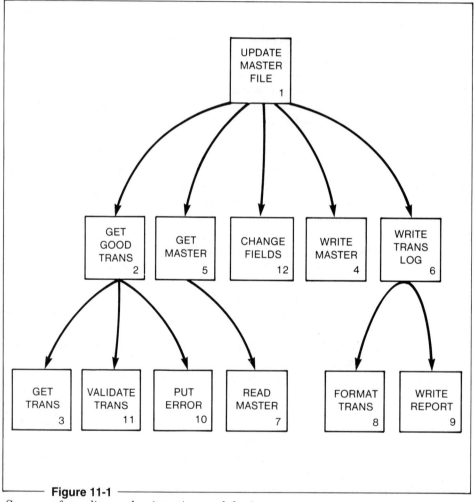

Figure 11-1

Sequence for coding and unit-testing modules in a program.

will have some control over how programs are written. By standardizing certain aspects of program code, the company ensures that the programs can be read and correctly interpreted by any programmer, not only by the author. Standards often describe

naming conventions for data items,

naming conventions for subroutines,

format and indentation for certain complex statements (such as IF–THEN–ELSE),

restrictions on the use of branching verbs (such as GOTO),

the organization of modules within a program.

Generally, adherance to programming standards results in understandable code. If your installation has programming standards, then that is another topic that should be considered in a code review: all programs should adhere to the established standards.

TESTING   As programmers code modules, they *unit-test* them—that is, they create data to pass to each module in order to observe its behavior. Before a program (or a module, or a system) is accepted for production (that is, used for real processing), it must be tested vigorously and thoroughly. *Acceptance testing,* as this is called, is generally the responsibility of someone other than the programmer. Programmers want to demonstrate that their programs work; acceptance testers want to find all the programs that do not meet standards, and prevent them from being released for production. Thus, the purpose of acceptance testing is to prove that a module does not work according to its specifications.

Consider the analogy of the Quality Control group at an automobile manufacturing plant. Their job is to find flaws in the cars before they roll off the assembly line and into the dealer's showcase. The Quality Control people do not waste their time looking for all the things about a new car that are *right;* they note only the things that are *wrong.* And because time is limited, they cannot scrutinize every minute detail; instead they concentrate on the areas that are most likely to have flaws: joints, hinges, curves, edges. They are on the lookout for damaged parts, streaks in the paint, moving parts that don't, stationary parts that aren't. This group is successful only if it finds flaws!

Other Quality Control groups focus on the car's performance. Their job is to push the car to its limits and show that it does not meet the standards set for it: that emergency systems malfunction, that the automobile does not accelerate properly, that the brakes pull to one side, that the electrical system cannot handle peak loads. All inspectors and testers are on the prowl for flaws, errors, and malfunctions. And so it is with acceptance testing.

Knowing this, the programmer can take steps to test modules and find errors (and correct them) before the acceptance test team gets its chance to do so.

Although every programmer would like to prove that a program works under all circumstances, the only way to do that is to test it with all possible pieces of input data (this is like scrutinizing every square centimeter of a new automobile with a microscope), which, it turns out, is an infinitely large set of test data. Rather than attempt the impossible, we should spend our limited time like the automobile inspectors: looking carefully at areas that are most likely to have problems, while only glancing at the rest. The philosophy is this: if the areas that are likely to have errors do, then we find them before the module leaves our hands; and if the areas most likely to have errors do not, then it is probable that the rest of the module works okay as well. Our goal is to design test data so we can observe how each module handles input data, how modules produce output, and how well modules do their jobs.

INPUT   For each module on the structure chart, examine the input data it receives from its boss. We are going to put different values in the input parameter and call the module to see how it processes this hand-picked data. If the results it returns are the same as the results we predicted, then we did not discover any errors. Conversely, if the module fails by either abending (terminating abnormally) or by returning incorrect results, then our test successfully uncovered an error that we can now correct and retest before the module is submitted for acceptance testing.

Test cases are made up of two parts: the test data itself, and the expected results (that is, what the module ought to return in its output parameters if it functions according to its module specifications). Both the test data and the expected results are written down for future reference. We will need this data when we perform the test and compare the actual results to the expected ones. We also need test data in the future, after the module is in production and may be changed. When we test a production program or module after it has been changed, we are doing *regression testing*. We want to be sure the changes have not adversely affected the module's performance. What better way than to use the original test data?

The valid values for an input parameter generally fall into four categories: they must be a certain *data type,* or within a certain *range,* or within a certain *set,* or there may be multiple *interdependent* parameters. We will look at an example of each type of input parameter, and create some sample test cases.

DATA TYPE   Some modules expect their input to belong to a certain type—for example numeric, alphabetic, or integer. Here are some examples.

Amount of purchase must be positive numeric.
Number of credits must be positive integer.
Client name must be alphabetic.
Funding-source account balance must be signed numeric.

When we test a module that is supposed to handle data of a certain type, we try to prove either that it handles valid values incorrectly, or that it processes invalid values as if they were valid, or that it abends (terminates abnormally) when passed invalid data. Therefore we create a test case for each of these conditions. We test it with both valid and invalid values in the input parameter. Figure 11-2 shows a sample module and the test cases we might use to prove that it does not work according to specifications.

SET Some modules expect the input parameter to contain one of a discrete set of values. It should correctly process all valid values and reject those values that are not part of the set. Here are some examples.

Bank transaction indicator must be either—

"DS"	Deposit to savings account
"WS"	Withdrawal from savings account
"DC"	Deposit to checking account
"CC"	Cash check
"TF"	Transfer funds from one account to another

Student class designation must be either "FR", "SO", "JR", or "SR".

Student's preferred reading mode must be either "B", "L", "T", or "P" (for braille, large print, tape, or regular print).

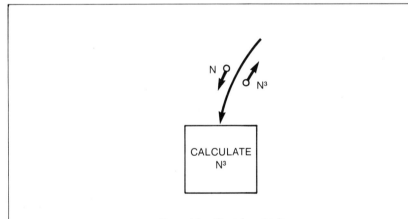

N must be signed numeric

TEST DATA	EXPECTED RESULTS	ACTUAL RESULTS
N = −9 N = +9 N = "B"	$N^3 = -729$ $N^3 = 729$ $N^3 = 0$	

Figure 11-2

Test cases for module expecting parameter of a data type.

The easiest way to prove that the module does not work is to create one test case for each valid value, and one for an invalid value. This set of test cases will reveal any processing errors due to omission (the programmer forgot to include one of the valid values), as well as the error of forgetting to reject invalid values. Consider this piece of faulty code that should handle four reading modes (B, L, T, and P), and *should reject* all other values:

```
IF MODE IS "B"
    DO B-ROUTINE
ELSE
IF MODE IS "L"
    DO L-ROUTINE
ELSE
IF MODE IS "T"
    DO T-ROUTINE
ELSE
    DO P-ROUTINE.
```

What happens when we pass this module a value of "X" in MODE? Figure 11-3 illustrates a module expecting data to be in a certain set, and the test cases we might construct for it.

RANGE   Some modules expect data values to be within a certain range—that is, between two values. The following are examples.

A client must be between 2 years 3 months and 4 years 6 months to qualify for a certain funding program.

Customer balances between $100 and $499 will be subject to a 3% service charge.

The number of drivers for an insurance policy must be between 1 and 8.

As it happens, errors in such modules usually cluster at the two ends of the range. Therefore, we focus our testing attention on the ends of the range. If there are errors, then we'll probably find them. Conversely, if the module handles both the high and the low values properly, it will probably also correctly handle the ones in between.

We create test cases for the lowest valid value in the range as well as the highest valid value. We also create test cases for the nearest invalid values to be sure they are rejected—that is, the invalid value just a little lower than the lowest valid value, and the invalid value just a little higher than the highest valid one (see Figure 11-4). Figure 11-5 shows an example of test cases we might create to test a module that expects input to be within a certain range.

INTERDEPENDENT PARAMETERS   Some modules receive multiple input parameters from the calling module. And sometimes the module processes different combinations of values differently. We want to test all the interesting combinations of input data (we cannot test *all* combinations, because that is an infinite set of test data), the combinations that the programmer is likely to have handled wrong. In the example in Figure 11-6, the module should return a value of 10.75% in INTEREST-RATE if PARENTS-INCOME is less than $11,000 and APPLICANT-STATUS is "Y"; in all other cases it should return an INTER-

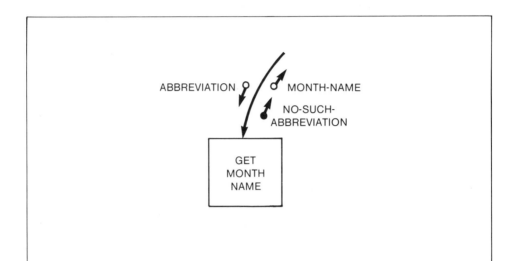

TEST-DATA	EXPECTED RESULTS		ACTUAL RESULTS
ABBREVIATION =	MONTH-NAME =	NO-SUCH-ABBREVIATION	
"JAN"	JANUARY	N	
"FEB"	FEBRUARY	N	
"MAR"	MARCH	N	
"APR"	APRIL	N	
"MAY"	MAY	N	
"JUN"	JUNE	N	
"JUL"	JULY	N	
"AUG"	AUGUST	N	
"SEP"	SEPTEMBER	N	
"OCT"	OCTOBER	N	
"NOV"	NOVEMBER	N	
"DEC"	DECEMBER	N	
"ABC"	——	Y	

Figure 11-3

Test cases for module expecting parameter of a set.

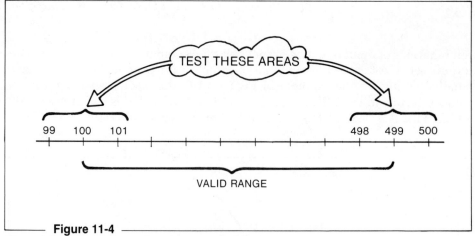

Figure 11-4

Potential error-causing values in a range of data.

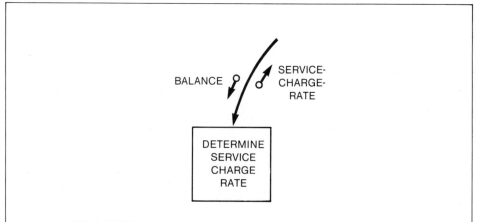

If BALANCE between $100 and $499, SERVICE-CHARGE-RATE = 3%.
Otherwise SERVICE-CHARGE-RATE = 0.

TEST DATA	EXPECTED RESULTS	ACTUAL RESULTS
BALANCE =	SERVICE-CHARGE-RATE =	
99	0	
100	0.03	
101	0.03	
498	0.03	
499	0.03	
500	0	

Figure 11-5

Test cases for module expecting parameter in a range.

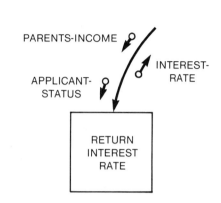

If PARENTS-INCOME < $11,000 and APPLICANT-STATUS = 'Y', INTEREST-RATE = 10¾%. Otherwise INTEREST-RATE = 14%.

TEST DATA		EXPECTED RESULTS	ACTUAL RESULTS
PARENTS-INCOME	APPLICANT-STATUS	INTEREST-RATE	INTEREST-RATE
10,999	Y	10.75%	
11,000	Y	14%	
10,999	N	14%	
11,000	N	14%	

Figure 11-6

Test cases for module processing interdependent parameters.

EST-RATE of 14%. Of course, we include a test for the one combination that should return a value of 10.75%. We also test the other three combinations that should return a value of 14% in order to discover if the programmer forgot to test combinations, and instead made the decision based on only one of the parameters.

A parameter may fall into several of the categories just discussed. One parameter might have to be numeric *and* within a certain range that depends on the value of another parameter. Naturally, this results in more test cases. The point to remember is that, because testing is time-consuming, we want to limit the number of test cases. And remember that, after the test is done, we must compare every actual result with every expected result. We do not want to wade through hundreds or thousands of test cases if we can avoid it. However, if you are unsure about the relative merit of a test case and wonder whether or not to include it, err on the side of testing too much and include it anyway. It just may be the one case that illustrates that the module does not work.

OUTPUT   For every module that produces printed output, the test is made with enough input to detect errors in page overflow. Also create test cases that cause the highest and lowest possible values to print out. Try to detect errors caused by defining output fields or edit patterns that are too short. Also test the printing of both positive and negative numbers.

For those modules that write output records on tape or disk, you must dump the file in order to examine its contents. Be sure to examine the first and last records: sometimes a programmer neglects to write the first record, or stops processing before writing the last record, or neglects to close the output file which, in some systems, means that the last block of data is not written onto the tape or disk. Also create enough test cases to cause block overflow.

OTHER TESTING IDEAS   Test modules that read input files with empty data sets (they encounter the end-of-file marker the first time through).

For table-searching modules, test not only the set of valid input parameters, but also the sequence in which valid and invalid input parameters are passed. By mixing up the order, we can detect errors due to neglecting to reset a subscript or pointer, or forgetting to reset the NOT-FOUND flag. You simply create test cases to cover the following combinations of calls:

HIT followed by HIT at a higher location in the table

HIT followed by HIT at a lower location in the table

HIT followed by NO HIT

NO HIT followed by HIT

NO HIT followed by NO HIT

In addition to the ideas suggested above, follow your own instincts. If you think of a test case that has some chance of detecting an error that would otherwise go unnoticed, include it. Be creative!

DATA

Some new systems use the existing data files or database or both. In many cases, however, the new system requires more data, or data stored on different devices, or data stored in a different format, or all of these. We must either *create* new files or databases by entering the data for the first time, or *convert* existing data to new formats or storage media.

File conversions are usually done by people in the data-processing department: computer operators or computer programmers use utility programs that essentially copy data from one file to another while rearranging some fields, deleting others altogether, merging several files into one, and so forth. The data is already available, even though it is currently stored on a different device or in a different format than the new system will use.

File creation means storing data in computer files for the first time. If you install a computer system where there was none before, you must create computer files from the data now kept in manila folders in file cabinets. Or if you replace a manual system with an automated one in a company that already has computers, you may need to add new files for the new functions.

File creation can be done by data-entry professionals who key the data from source documents onto punched cards, magnetic tape, diskette, or other computer-readable media. The source documents generally are prepared by the user departments. They dig through the original documents and copy the data onto source documents that the data-entry people use. In other cases, file creation is handled completely by the user department, especially when data can be entered by means of a terminal. Special programs must be written to display prompts and templates, and data can be validated as soon as it is entered.

Databases are more complex than other data files because they contain not only data fields, but also information about data relationships and much overhead data. They are more difficult to create and to maintain. Database-management systems (DBMSs) include both the language for creating the database (called the data-definition language, or DDL) and the language for referencing data in the database (called the data-manipulation language, or DML). We mentioned in Chapter 8 that the Database Administrator (DBA) is responsible for the database, including both its creation and its maintenance. Thus it is the DBA who makes the decisions related to database creation. The DBA translates the logical database design into the physical database format appropriate for the DBMS, and then uses the DDL to create the database (or oversees its creation).

No matter who actually creates or converts the data for the new system, it must be converted (and file contents verified) by the time the system components are integrated and tested as a whole unit. We should have live data available for certain system tests, such as those designed to test the speed of the system, or to test its ability to handle large volumes of data. (System testing is discussed in greater length in another section of this chapter.) We do not usually use live data for the module testing discussed earlier for two reasons: live data is usually "clean," that is, it does not contain the rare cases we must test for; and there is usually too much data to wade through.

PROCEDURES

We established procedures for everything during the design stage; what remains to be done is to write the reference manuals and operations documentation.

Ideally, someone who is good with words *and* who understands both the system and the reader's point of view will write the reference manuals. Some companies hire technical writers to do this very important job. Others pair off a writer with a system person and have them produce the manuals as a team. Poor documentation usually is produced either by a skilled, knowledgeable data processor who is a poor writer, or by a good writer who does not understand the system being documented. Assuming you will someday be faced with this task, consider the following guidelines.

Both the user and the operations manuals are the physical embodiment of the procedures established already. The contents of the manuals follow the outlines we produced during the design stage. The groundwork has been done: now we simply build upon it.

Choose a writing style you feel comfortable with. Do not try to write like someone else. Following the outline, write your first draft as it comes to you. After writing a first draft, go back and critique it. Is it complete? Are any sections redundant? Does the organization of the written material match the outline? Are there any instructions or procedures that need further explanation or clarification? Revise it where necessary.

Once convinced that the document contains exactly the right content, go back and critique it again, this time with an eye toward making it concise. Weed out unnecessary verbiage, deleting anything that does not contribute to the reader's understanding. Look for phrases that can be replaced by single words, and for multisyllabic words that can be replaced by simple ones. Your reader should not need a dictionary to read the reference manual. Remember that people generally resort to reading manuals only after they have given up trying to figure something out for themselves. They are probably already frustrated: do not add to their frustration by overpowering them with writing that should appear only in a professional journal! Revise it again.

Now go back again and check the document for both technical accuracy and style. Do illustrations and references to them in the text match? Are any mislabeled? Do titles of chapters, units, and sections state precisely what the text covers? In procedure descriptions, be sure you make use of imperative statements rather than the passive voice. Here is an example of imperative: "Lift the printer cover by the handle until it clicks into place; press down on the printer-mechanism release lever, exposing the print ribbon." Here is an example of passive voice: "The printer cover must be lifted by the handle until a click is heard; the printer-mechanism release lever then is pressed down, and the print ribbon is exposed." Remember, you are telling the reader how to *do* something. Revise this draft too.

Finally, go through your revised document one more time, examining your use of illustrations. Some will be unnecessary (pull them out), but often you will want to add pictures to accompany your "thousand words." Line drawings are better than photographs, because they omit irrelevant details and highlight important areas to which you want to draw the reader's attention. Some illustrations you may want to include (see Figure 11-7) are

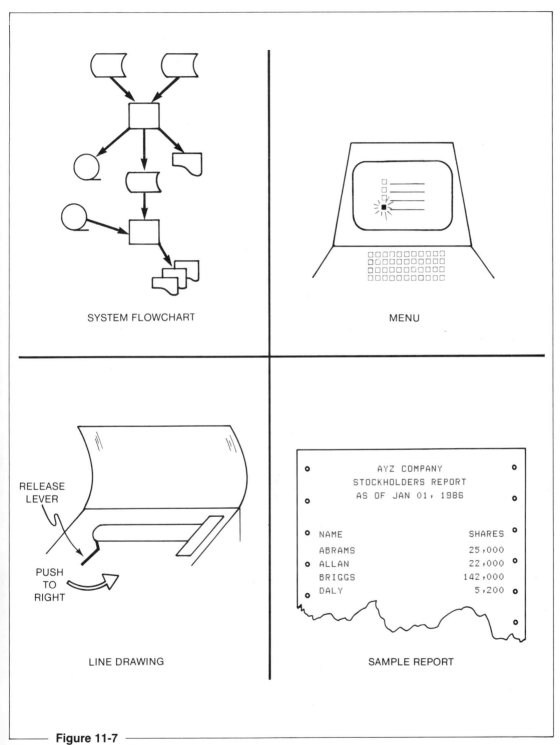

Figure 11-7

Illustrations found in documentation.

system flowcharts,

screen displays,

line drawings of pieces of equipment with parts identified,

sample output reports.

When arranging the final text, try to place illustrations and accompanying text on the same page or on facing pages, so the reader can refer to both without turning pages.

All manuals should be *tested* before they are released for use. Operators and users should review the material before it goes into print. If there are still problems, questions, ambiguities, or mistakes, then we have one last chance to correct them.

TRAINING PROGRAMS

During implementation, we produce training materials and deliver training programs. Many companies enlist the service of professional educators for this crucial step. In Chapter 10, we discussed the differences between self-paced and directed training programs; we also examined various media used in self-study programs, including videotaped presentations, programmed texts, and computer-based training. Finally, we looked at other pedagogical approaches including classroom learning and on-the-job training. Each of the approaches requires different training materials, ranging in sophistication from a brief outline of topics for one-on-one hands-on training to integrated multimedia materials. The production of such materials and the delivery of both user and operations training must be done by professionals if it is to be of the highest quality.

The training itself can be scheduled many ways. If the system is being released in phases, or versions, then train the users and operators only in the areas they need to know in order to use the next release of the system. Then while they are actually using that version of the system on the job, they can be receiving the training they will need in order to use the next release. This allows us to overlap training and use of the system (See Figure 11-8). It also allows the people to become comfortable with some system functions before they are faced with learning more.

For those systems that are released all at once, training can be done while the system is being *parallel-tested*—that is, while it is running at the same time as the system being replaced, so that results of both systems can be compared and any bugs worked out before the old system is abandoned altogether. (More on parallel testing later.)

Training can also be done before the new system is delivered, either at another site where the system is already installed and running, or on a simulated system.

We run training programs either at the user's site or in a centralized training area. If we train the user (or the operator) at their location, then a trainer teaches a few people who will later pass on what they know to others in the same department. Or the people to be trained can come to a training site where all the materials they need, as well as the training staff, are readily available.

For self-paced training, the materials we produce are all that the user or operator will have when learning. Therefore they must be self-explanatory, clear, concise, and well organized. They should present frequently used material at the beginning of training, and should hold off on esoteric functions until the end.

Remember that the trainee wants to learn *how* to use or operate the system. Focus on the objectives established during design.

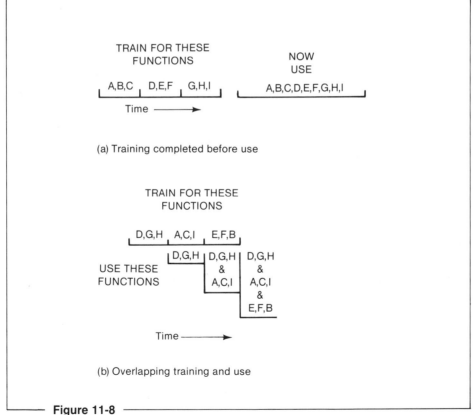

Figure 11-8

Delivering all training before using system vs overlapping training with use of parts of system.

SYSTEM TESTING

Theoretically, all the individual pieces of the system now work—the problem is that each piece works independently, but in real life all the pieces have to work together. Therefore we must perform a series of *system tests*. This involves the integration of all system components in a test environment. Some interfaces have already been tested—for example, interfaces between programs. But during system testing, we put all the pieces together and deliberately subject the system to various tests to determine whether or not the whole system meets the user's specified requirements. This is our last chance to detect and correct errors before the system is installed for acceptance testing. Therefore we want our tests to really push the system to its limits. And although we cannot anticipate every condition, environment, or situation in which the system will ultimately be expected to perform, there are some typical errors on which we can concentrate.

Who should be involved in system testing? Another way to ask that question is: Who should *not* be involved in system testing? The team that developed the system is not likely to be as objective in testing the system as an independent group. Only one or two of the system analysts who worked on the project should be included in this step. Of course, users become far more involved in system testing than they were in unit testing.

We said earlier that the goal of module testing is to demonstrate that a module does not perform the way its specification says it ought to. Similarly, in system testing, our goal is to demonstrate that the system does not perform the way the system specification says it ought to. The guide to system testing is the system specification, because it contains descriptions of all required functions and features of the system, as well as constraints.

SYSTEM TEST CASES

There is a wide variety of tests to which we wish to subject the system; this section will touch upon some of the major ones. These topics and others are discussed by Myers.*

VOLUME   The purpose of volume testing is to demonstrate that the system cannot handle as much data as the system specification said it would. We subject the system to a large number of records (several reels of transaction tapes, for example) in an attempt to abend the system or show that it incorrectly processes data. The user is usually called upon to provide test data for volume testing.

STRESS   The purpose of stress testing is to prove that the system malfunctions under peak loads. We subject the system to large quantities of data over

*Myers, G. *The Art of Software Testing.* Wiley, 1979.

a short period of time (as opposed to volume testing, where time is not of the essence). Stress testing is particularly important for online systems in which activity takes place in spurts.

USABILITY   The purpose of usability testing is to demonstrate that the system is not user-friendly. This test involves the examination of output messages, error-handling procedures, normal operating procedures, and so forth. Of course, the user must perform this test if it is to be meaningful.

SECURITY   The purpose of security testing is to demonstrate that data and programs can be stolen, copied, destroyed, or altered without authorization; that unauthorized data can be accepted as input; that output can be delivered to unauthorized recipients; and so on.

PERFORMANCE   The purpose of performance testing is to demonstrate that the system is unable to process data as quickly as the system specification says it will. This includes response time in online systems as well as total processing time in batch systems.

DOCUMENTATION   The purpose of documentation testing is to demonstrate that user or operations documentation is inaccurate or incomplete. The user (or operator) tries to use the system with only the documentation as a guide, testing all the examples and illustrations to determine whether or not they match reality.

PROCEDURES   The purpose of procedure testing is to prove that procedures are cumbersome, unwieldy, confusing, or wrong. The user, once again, is the only one who can objectively perform this test. Having included the user in the design of both procedures and their documentation should eliminate many potential problems at this stage of system testing.

ACCEPTANCE TESTING

The acid test for any system is *acceptance testing*. During this step, the developers step out of the picture completely, and the user takes over. The system is installed and run with live data under real (or sometimes simulated) working conditions by actual users.

A user might take a variety of approaches to acceptance testing. One popular approach is called *parallel* testing: the new and old systems are run using the same data for a period of time (for example, for three months). During that time, results from the two systems are compared. Errors detected in the new system are corrected (the developers must participate in the correcting, but not in the testing). When the user determines that the new system satisfies the stated requirements, the old system can be retired and the new one accepted for production.

Another approach is to introduce and test the new system in *phases,* or *versions.* Each version of the system is run in a real environment, usually in parallel with the old one, and fine-tuned. As each phase is accepted, a new one with more functions is added and tested in the same way. This continues until the entire system is in place.

Still another approach is to put the not-yet-accepted system directly into production, immediately replacing the old one. Bugs are worked out as they surface. As you might have guessed, this radical approach is one of the riskiest ways to do acceptance testing.

SYSTEM EVALUATION AND USER FEEDBACK

No system-development project ever really "ends." Rather, the end of one so-called project simply marks the beginning of the next one. A truly useful system evolves to meet changing user needs. Therefore it is vital to the success of the system to get the user's evaluation of it.

Undoubtedly, the user will require changes to the system, sometimes as soon as it is put into production. In many cases, the new system contains only the most vital features the user wanted in the first place. In the interest of time and money, the user prioritized all known needs and selected only some of them for inclusion in the system specification. Now is the user's chance to ask for more (or more sophisticated) features. As data-processing people, we may become aware of new technology, such as new hardware or software, that could improve the system's usability or operating cost.

We must actively seek user feedback. Customer representatives or marketing personnel perform this function for software companies. They send questionnaires, make telephone calls, and visit the installation looking for feedback. Companies that support an in-house data-processing staff can do almost the same thing. Some companies have user department/data-processing department liaison people who bridge the gap between the two departments. It is likely that the analyst who worked on the system project has developed enough of a relationship with the user to ask for, and get, honest feedback on the system.

The point is this: Someone must ask questions about the user's reaction to the system. Someone must ask what the user's new needs are. The answers to these questions may well mark the beginning of a new system-development project.

SUMMARY

The system-development process culminates with the actual building and integration of the various system components. People with special skills carry out the plans developed during the design stage. The various individuals follow an implementation plan that is based on scheduled delivery dates and acceptance standards for every product. Frequent reviews of progress compared to the schedule help detect problems while there might be time to solve them without affecting the rest of the schedule. Sometimes deadlines are missed, and the implementation schedule for subsequent steps must be adjusted accordingly.

Each component is built, installed, and tested independently: this is called unit testing. Technicians install and test hardware. Users, programmers, operators, and database specialists fill files and databases with application data. Programmers write and test software; commercial packages are installed and tested. Technical writers draft reference manuals and other documentation, while trainers teach the users and operators how to interface with the new system.

The components are then integrated, and system testing takes place. The user is more actively involved in system testing than in unit testing. The system is subjected to a variety of tests designed to determine whether or not it satisfies the requirements stated in the system specification. After some fine-tuning and adjusting, the new system goes into acceptance testing.

During acceptance testing, the user takes over completely. The new system might be run in parallel with the old one; it might be released and tested in versions; or it might be installed for production immediately (gulp!).

Once in production, the system can be evaluated by users. Changes the users would like to see, new needs to satisfy, and introduction of new technology are all topics that can become the jumping-off point for a new system-development project.

KEY WORDS

Acceptance criteria Minimum standards a product must meet in order to be considered complete.

Acceptance test A test designed to determine whether or not a system will be accepted for production. It usually refers to test of an entire system, but can also refer to a module or a program.

Diagnostic test A type of test run on computer hardware in order to detect hardware errors.

Parallel testing Running an old system and a new system at the same time and with the same data, and comparing the results in order to identify discrepancies. This is a method of doing system acceptance testing.

Programming standards Rules for programming to which programmers are expected to adhere.

Regression testing Repeating a test in order to determine if a change made to the product being tested has adversely affected the product.

Stub A not-yet-completed module. It takes the place of a real module for the purpose of testing its boss.

System test A test designed to demonstrate that the system does not perform according to its specification.

Unit test An independent test of one small part of a system, such as a module, a program, or a procedure.

QUESTIONS

1. Why is an implementation plan so important during the final system-development step?

2. Why are frequent reviews of progress compared to schedule so important?

3. What is involved in preparing a site for the installation of computer equipment?

4. How would you find out what environmental changes you would have to make before installing computer equipment?

5. Why is the structure chart the best guide for coding and testing programs?

6. How could the structure chart be used to distribute the programming task among a group of programmers?

7. Many systems are coded from the top down. Why would we sometimes want to code from the bottom up?

8. What is the user's involvement in system testing?

9. What is the user's involvement in acceptance testing?

10. Describe three approaches to acceptance testing.

11. Why must time for error detection and correction be built into the implementation plan?

12. What is the difference between an estimate and a promise?

13. Why is it important to seek user feedback on a system that has been released for production?

System Specification Documents for Resource Center

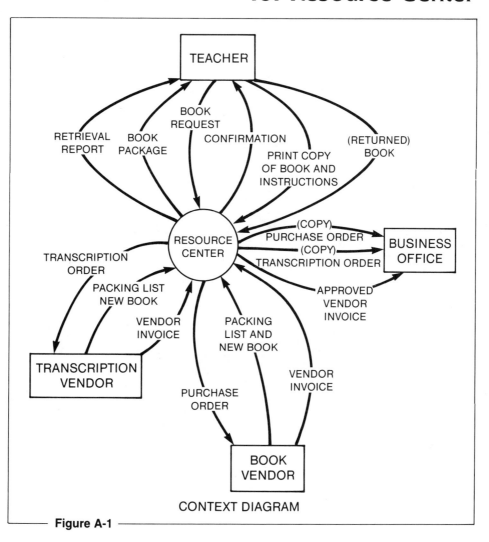

CONTEXT DIAGRAM

Figure A-1

308

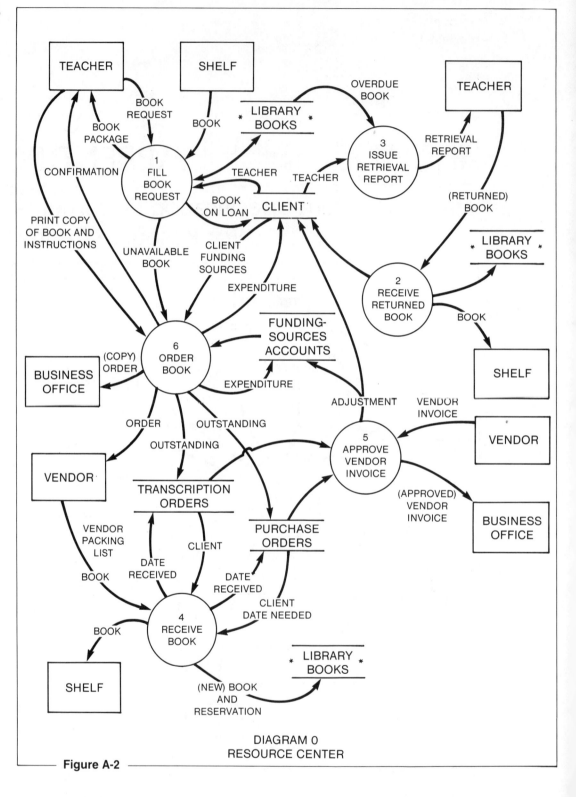

DIAGRAM 0
RESOURCE CENTER

Figure A-2

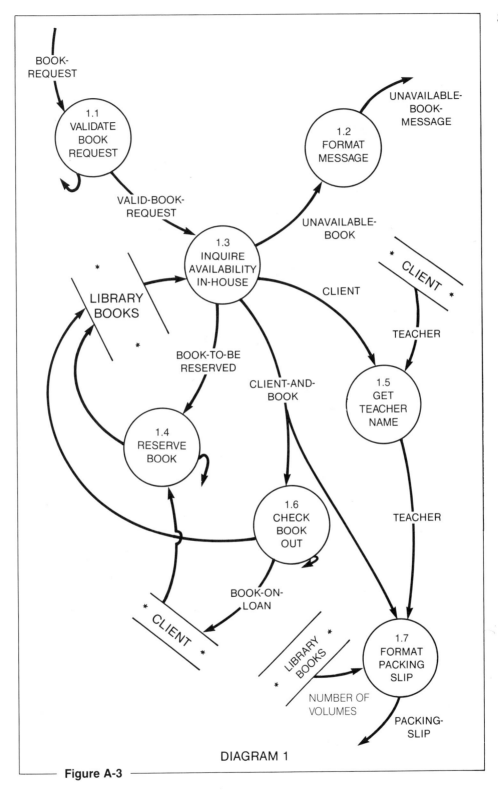

DIAGRAM 1

Figure A-3

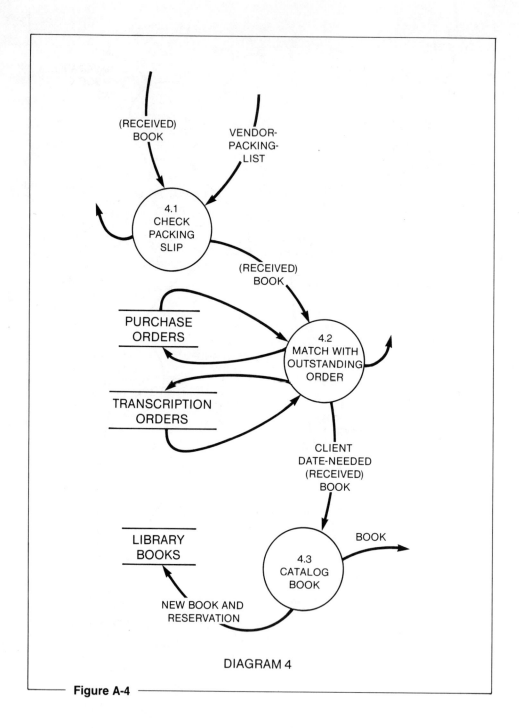

DIAGRAM 4

Figure A-4

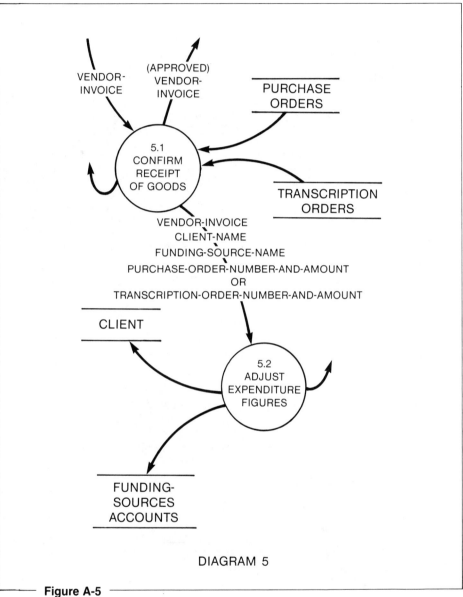

DIAGRAM 5

Figure A-5

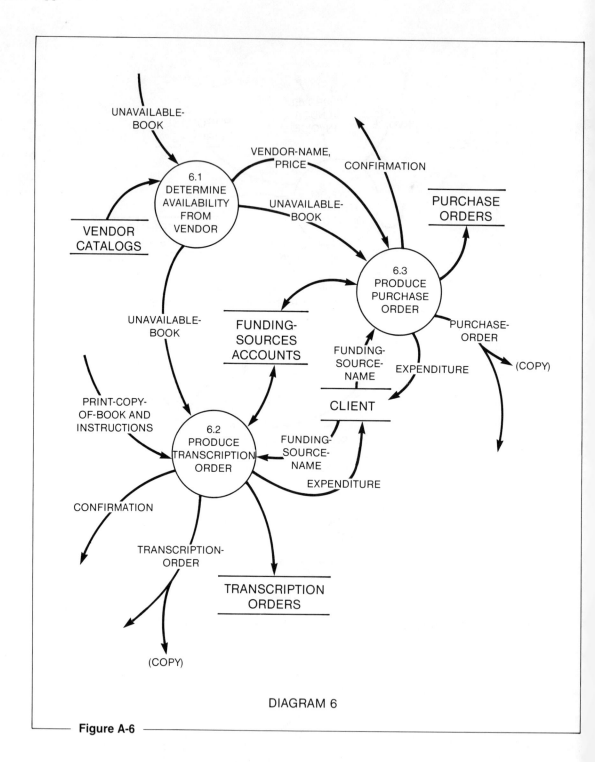

DIAGRAM 6

Figure A-6

ADJUSTMENT = Client-name + Funding-Source + [Purchase-Order-Number | Transcription-Order-Number] + Adjustment-Amount

BOOK-ON-LOAN = Catalog-Number + Title + Date-Due

BOOK-PACKAGE = Book + Packing-Slip

BOOK-REQUEST = Teacher-Name + Client-Name + Date-Needed + Title + Author + Copyright + Medium + ['Needed-Now' | 'Reserve']

CONFIRMATION = Teacher-Name + Client-Name + Title + Medium + Date-Ordered + Todays-Date

CLIENT FILE = {Client-Record}

CLIENT-RECORD = Client-Name + Date-Of-Birth + Address + Teacher-Name + {Reading-Mode} + {Book-On-Loan} + {Expenditure} + {Funding-Source-Name}

CURRENT-BORROWER = Client-Name

CURRENT-STATUS = ['In' | 'Out']

EXPENDITURE = Funding-Source-Name + Date-Ordered + Description + Amount + [Purchase-Order-Number | Transcription-Order-Number] + Purchase-Type

EXPENDITURE-APPROVED = Client-Name + Date-Ordered + Description + Amount + [Purchase-Order-Number | Transcription-Order- Number] + Purchase-Type

FUNDING-SOURCE-CODE = ['P' | '9' | 'S' | 'A' | 'M']

FUNDING-SOURCE-NAME = ['Preschool' | 'Title IX' | 'SSI' | 'APH' | 'Multihandicapped']

FUNDING-SOURCES ACCOUNTS FILE = {Funding-Sources-Account-Record}

FUNDING-SOURCES-ACCOUNT-RECORD = Funding-Source-Name + Beginning-Balance + Total-Expenditures-Approved + Current-Balance + {Expenditure-Approved}

LIBRARY BOOKS FILE = {Library-Book-Record}

LIBRARY-BOOK-RECORD = Title + Author + Copyright + Publisher + Catalog-Number + Current-Status + Current-Borrower + (Reservation) + Medium + Number-Of-Volumes + Date-Out + Date-Due

MEDIUM = alias Reading-Mode

ORDER = [Purchase-Order-Form | Transcription-Order-Form]

OVERDUE-BOOK = Current-Borrower + Title + Date-Due

PACKING-SLIP = Title + Catalog-Number + Medium + Teacher-Name + Client-Name + Date-Needed + Number-Of-Volumes

PURCHASE ORDERS FILE = {Purchase-Order-Record}

PURCHASE-ORDER-RECORD = Purchase-Order-Number + Vendor-Name + Date-Ordered + Date-Needed + Description + Amount + Client-Name + Funding-Source-Name + Date-Received

PURCHASE-ORDER-FORM = Purchase-Order-Number /*preprinted*/ + Vendor-Name + Vendor-Address + Vendor-City + Vendor-State + Vendor-Zip-Code + Date-Ordered + Date-Needed + Description-Of-Goods-Or-Services + Amount + Funding-Source-Code

Figure A-7

Data Dictionary.

PURCHASE-TYPE = ['Braille Book' | 'Large Print Book' | 'Cassette' |
    'Braille Transcription' | 'Large Print Transcription' | 'Cassette
    Transcription']
READING-MODE = ['Braille' | 'Large Print' | 'Regular Print' | 'Cassette']
RECEIVED-BOOK = Book + [Transcription-Order-Number | Purchase-
    Order-Number]
RESERVATION = Client-Name + Date-Reservation-Made + Date-
    Needed
RETRIEVAL-REPORT = Teacher-Name + Client-Name + {Title + Date-
    Due}
TRANSCRIPTION ORDER FILE = {Transcription-Order-Record}
TRANSCRIPTION-ORDER-RECORD = <u>Transcription-Order-Number</u> +
    Vendor-Name + Date-Ordered + Date-Needed + Title + Medium +
    Amount + Client-Name + Funding-Source-Name + Date-Received
TRANSCRIPTION-ORDER-FORM = Transcription-Order-Number
    /*preprinted*/ + Vendor-Name + Vendor-Address + Vendor-City +
    Vendor-State + Vendor-Zip-Code + Title + Medium + Amount +
    Client-Name + Teacher-Name + Funding-Source-Code
UNAVAILABLE-BOOK = Book-Request
VENDOR-INVOICE = [Purchase-Order-Number | Transcription-Order-
    Number] + Vendor-Name + Invoice-Amount + Invoice-Date

Figure A-7 *(continued)*

1.1   VALIDATE BOOK REQUEST

Field	Validation Criteria
TEACHER-NAME	Nonblank
CLIENT-NAME	Nonblank
DATE-NEEDED	mm dd yyyy
TITLE	Nonblank
MEDIUM	['BRAILLE'	'LARGE PRINT'	'CASSETTE']

1.2   FORMAT MESSAGE

Title   Medium   'UNAVAILABLE'

1.3   INQUIRE AVAILABILITY IN-HOUSE

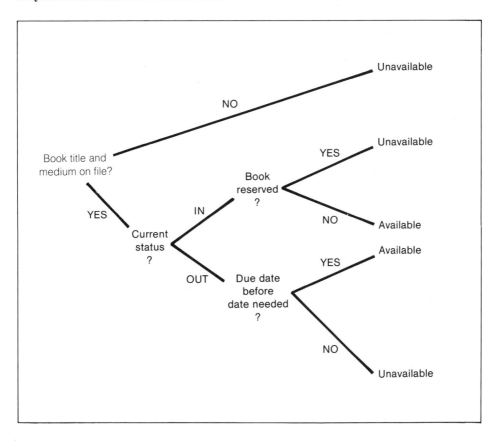

Figure A-8

Process specification.

1.4   RESERVE BOOK

Using TITLE
      retrieve LIBRARY-BOOK-RECORD from LIBRARY BOOKS FILE
Using CLIENT-NAME
      retrieve CLIENT-RECORD from CLIENT FILE
      If client not on file reject RESERVATION
      Copy CLIENT-NAME, DATE-NEEDED, and todays date into RESERVATION

1.5   GET TEACHER NAME

Using CLIENT-NAME
      retrieve TEACHER-NAME from CLIENT-RECORD

1.6   CHECK BOOK OUT

Using TITLE
      retrieve LIBRARY-BOOK-RECORD from LIBRARY BOOKS FILE
      Change CURRENT-STATUS to 'OUT"
      Copy CLIENT-NAME into CURRENT-BORROWER
      Copy todays date into DATE-OUT
      Set DATE-DUE = todays date + 6 months
Using CLIENT-NAME
      retrieve CLIENT-RECORD from CLIENT FILE
      Copy TITLE, CATALOG-NUMBER, DATE-DUE into next available BOOK-ON-LOAN

1.7   FORMAT PACKING SLIP

Using TITLE
      retrieve NUMBER-OF-VOLUMES from LIBRARY BOOKS FILE
Complete PACKING-SLIP

2   RECEIVE RETURNED BOOK

Using CATALOG-NUMBER of returned book
      retrieve LIBRARY-BOOK-RECORD from LIBRARY BOOKS FILE
Change CURRENT-STATUS to 'IN'
Using CURRENT-BORROWER
      retrieve CLIENT-RECORD from CLIENT FILE
      Delete BOOK-ON-LOAN entry for this book TITLE

3   ISSUE RETRIEVAL REPORT

For each LIBRARY-BOOK-RECORD whose DATE-DUE is greater than todays date
      Extract TITLE, CURRENT-BORROWER, and DATE-DUE
Using CURRENT-BORROWER
      retrieve CLIENT-RECORD from CLIENT FILE
Complete RETRIEVAL-REPORT

Figure A-8 *(continued)*

4.1 CHECK PACKING SLIP

Compare contents of BOOK-PACKAGE with list of contents on PACKING-SLIP
If there are any discrepancies
    Notify Librarian
    otherwise mark PACKING-SLIP 'OK' and initial it

4.2 MATCH WITH OUTSTANDING ORDER

For TRANSCRIPTION-ORDER:
    Using TRANSCRIPTION-ORDER-NUMBER
        retrieve TRANSCRIPTION-ORDER-RECORD from TRANSCRIPTION ORDERS FILE
For PURCHASE-ORDER:
    Using PURCHASE-ORDER-NUMBER
        retrieve PURCHASE-ORDER-RECORD from PURCHASE ORDERS FILE
If matching order number cannot be found
    Notify Librarian
    Otherwise continue
Copy todays date into DATE-RECEIVED

4.3 CATALOG BOOK

Assign unique CATALOG-NUMBER to book
Mark all volumes of book or cassette tapes with CATALOG-NUMBER and volume number
Copy CLIENT-NAME, DATE-NEEDED and todays date into RESERVATION
Add LIBRARY-BOOK-RECORD to LIBRARY BOOKS FILE
Shelve all volumes of book

5.1 CONFIRM RECEIPT OF BOOK

Using PURCHASE-ORDER-NUMBER or TRANSCRIPTION-ORDER-NUMBER
    retrieve order from PURCHASE ORDERS FILE or TRANSCRIPTION ORDERS FILE
If there is no matching order
    Reject VENDOR-INVOICE, notify Librarian
If DATE-RECEIVED in order is blank
    Reject VENDOR-INVOICE, notify Librarian
Otherwise
    Mark VENDOR-INVOICE 'OK TO PAY'

5.2 ADJUST EXPENDITURE FIGURES

For VENDOR-INVOICE for PURCHASE-ORDER:
    Case 1: AMOUNT of VENDOR-INVOICE = AMOUNT of PURCHASE-ORDER-RECORD
        No adjustments needed
    Case 2: AMOUNT of VENDOR-INVOICE is less than AMOUNT of
            PURCHASE-ORDER-RECORD
    Difference = AMOUNT of PURCHASE-ORDER-RECORD minus AMOUNT of
    VENDOR-INVOICE
    Using FUNDING-SOURCE-NAME
        retrieve FUNDING-SOURCE-ACCOUNT-RECORD from FUNDING-SOURCES
        ACCOUNTS FILE
        Add difference to CURRENT-BALANCE
        Locate EXPENDITURE-APPROVED for this PURCHASE-ORDER-NUMBER
            Subtract difference from AMOUNT of EXPENDITURE-APPROVED

Figure A-8 *(continued)*

Using CLIENT-NAME
    retrieve CLIENT-NAME from CLIENT FILE
    Locate EXPENDITURE for this PURCHASE-ORDER-NUMBER
    Subtract difference from AMOUNT of EXPENDITURE
Case 3: AMOUNT of VENDOR-INVOICE is greater than AMOUNT of
          PURCHASE-ORDER-RECORD
Reject VENDOR-INVOICE
Notify Librarian
For VENDOR-INVOICE for TRANSCRIPTION-ORDER:
Case 1: AMOUNT of VENDOR-INVOICE = AMOUNT of
          TRANSCRIPTION-ORDER-RECORD
No adjustments needed
Case 2: AMOUNT of VENDOR-INVOICE greater than AMOUNT of
          TRANSCRIPTION-ORDER-RECORD
Difference = AMOUNT of VENDOR-INVOICE minus AMOUNT of
TRANSCRIPTION-ORDER-RECORD
Using FUNDING-SOURCE-NAME
    retrieve FUNDING-SOURCES-ACCOUNT-RECORD from FUNDING-SOURCES
    ACCOUNTS FILE
    Subtract difference from CURRENT-BALANCE
    Locate EXPENDITURE-APPROVED for this
    TRANSCRIPTION-ORDER-NUMBER
        Add difference to AMOUNT of EXPENDITURE-APPROVED
Using CLIENT-NAME
    retrieve CLIENT-RECORD from CLIENT FILE
    Locate EXPENDITURE for this TRANSCRIPTION-ORDER-NUMBER
        Add difference to AMOUNT of EXPENDITURE
Case 3: AMOUNT of VENDOR-INVOICE is less than AMOUNT of
          TRANSCRIPTION-ORDER-RECORD
Difference = AMOUNT of TRANSCRIPTION-ORDER-RECORD minus AMOUNT of
VENDOR-INVOICE
Using FUNDING-SOURCE-NAME
    retrieve FUNDING-SOURCES-ACCOUNT-RECORD from FUNDING-SOURCES
    ACCOUNTS FILE
    Add difference to CURRENT-BALANCE
    Locate EXPENDITURE-APPROVED for this
    TRANSCRIPTION-ORDER-NUMBER
        Subtract difference from AMOUNT of EXPENDITURE-APPROVED
Using CLIENT-NAME
    retrieve CLIENT-RECORD from CLIENT FILE
    Locate EXPENDITURE for this TRANSCRIPTION-ORDER-NUMBER
        Subtract difference from AMOUNT of EXPENDITURE

6.1 DETERMINE AVAILABILITY FROM VENDOR

Using book TITLE and MEDIUM
    Search vendor catalogs for book in client's reading medium
    Write down vendor prices

Figure A-8 *(continued)*

6.2   PRODUCE TRANSCRIPTION ORDER

Estimate AMOUNT at $100
Using CLIENT-NAME
    retrieve CLIENT-RECORD from CLIENT FILE
For each funding source for which client is eligible
    Using FUNDING-SOURCE-NAME
        retrieve FUNDING-SOURCES-ACCOUNT-RECORD from FUNDING-SOURCES
        ACCOUNTS FILE
        Select most restrictive funding source for which CURRENT-BALANCE
        is not less than AMOUNT
        Subtract AMOUNT from CURRENT-BALANCE for this funding source
Build EXPENDITURE for this order and add it to CLIENT-RECORD
Complete TRANSCRIPTION-ORDER-FORM
Get from teacher 1 or 2 copies of PRINT COPY of BOOK and INSTRUCTIONS
Write CONFIRMATION
Format TRANSCRIPTION-ORDER-RECORD and add it to TRANSCRIPTION ORDERS FILE
Package PRINT COPIES of BOOK, INSTRUCTIONS, and TRANSCRIPTION-ORDER-FORM
    and mail to transcription vendor

6.3   PRODUCE PURCHASE ORDER

Using CLIENT-NAME
    retrieve CLIENT-RECORD from CLIENT FILE
    For each funding source for which client is eligible
        Using FUNDING-SOURCE-NAME
          retrieve FUNDING-SOURCES-ACCOUNT-RECORD from FUNDING-SOURCES
          ACCOUNTS FILE
    Select funding source as follows:
        Case 1:   Client is eligible for APH
              and item is available through APH
              and APH CURRENT-BALANCE is not less than PRICE
          Set VENDOR-NAME = 'APH'
          Set FUNDING-SOURCE-NAME = 'APH'
        Case 2:   Otherwise
          Set VENDOR-NAME = least expensive vendor
          Set FUNDING-SOURCE-NAME = most restrictive funding source
          whose balance is not less than PRICE
Build EXPENDITURE and add it to CLIENT-RECORD
Build EXPENDITURE-APPROVED and add it to FUNDING-SOURCES-ACCOUNT-RECORD
Subtract AMOUNT of EXPENDITURE-APPROVED from CURRENT-BALANCE
Write CONFIRMATION
Write PURCHASE-ORDER-FORM
Format PURCHASE-ORDER-RECORD and add it to PURCHASE ORDERS FILE

Figure A-8 *(continued)*

Index